European Spatial Data for Coastal and Marine Remote Sensing

Editorial for Special Issue: European Spatial Data for Coastal and Marine Remote Sensing

Within the framework of the European Copernicus program, satellite data are effortlessly accessible to the scientific community, companies, and public. At the same time, the massive availability of digital data in various forms (spatialized or not) and across different fields opens new avenues of research and applications. Recent advances in the radiometric, spatial, temporal, and spectral resolution of sensors, as well as new data processing approaches in remote sensing, provide an extremely valuable set of tools for monitoring and mapping the various characteristics of marine, coastal, and aquatic systems (macroalgae, seagrasses, temperate reefs, coral reefs, salt marshes, mangroves, peatlands, aquatic vegetation, river systems, estuaries, deltas, etc.). It is crucial to increase coastal monitoring efforts in light of the growing number of studies highlighting the impacts of climate change, and in particular the sea-level rise that will radically redefine the physical state of the nearshore zone in the twenty-first century [2, 4]. In addition to climatic factors, there are also constraints on marine services imposed by the growth of coastal human populations, whose number is expected to nearly double by mid-century [1]. In conjunction with population growth, the exploitation of coastal resources has produced increasingly dramatic changes to coastal and inland aquatic habitats over the past 100 years [3].

In this special issue, 11 long papers, 6 short papers of original research and 3 abstracts of keyspeakers were published on topics covering a wide range of remote sensing applications, including marine, coastal, and aquatic ecosystems' monitoring and biodiversity management, coastal wetlands, nearshore land cover changes, deep learning and artificial intelligence innovations, marine, coastal and aquatic geomorphology, and coastal risk management. This special issue also provides new methodologies for optimizing the combined use of satellite/airborne data and field measurements for a comprehensive approach. Using satellite images of different spatial and temporal scales and applying innovative methods, several different types of coastal environments (bays, estuaries, sandy, and muddy systems) and several key parameters (vegetation, wind, etc.) and morphologies (shorelines, mud banks, wetlands) were analyzed and studied.

Harid et al. identify coccolithophores blooms of *Holococcolithophora sphaeroidea* in Algiers Bay from in situ and satellite observations (Modis data) over a period of 16 years (2003–2018). For the processing and analysis of the satellite images, a remote sensing reflectance index for coccolithophores (Cocco-Index) is used. The Cocco-Index proposed in this paper is based on a remote sensing reflectance (RSR) threshold between the green (531 nm, 547 nm, and 555 nm) and blue (443 nm, 469 nm, and 488 nm) bands of the MODIS sensor. This topic would benefit from further studies on the interactions between biological, physical, chemical, and environmental factors in the coccolithophore bloom areas.

Collin et al. investigate continuous mapping of salt marshes based on the normalized difference vegetation index (NDVI) ranges at three spatial and two temporal scales using Sentinel-2 (10 m), Dove (3 m), and unmanned aerial vehicle (UAV) (0.03 m) images. The NDVI ranges of the Sentinel-2 and Dove images were tracked with a five- and four-year lag, respectively. The NIR band of the UAV imagery was, therefore, modeled (R2NIR = 0.98) using a three-layer neutral network (NN) prediction based on red, green, and blue reflectance imagery, which in turn was calibrated/validated/tested by the bands of the Dove imagery. The 100-fold increase in pixel size allowed for the detection of decimeter-scale salt marsh and tidal flat objects.

Mury et al. investigate the contribution of near-infrared (NIR) and mid-infrared (MIR) bands from multispectral UAV and WorldView-3 (WV-3) super-spectral (SS) satellite imagery for mapping coastal ecomorphological features using two supervised classification algorithms, Maximum likelihood (ML) and support vector machine (SVM). They propose various combinations of spectral bands, visible + NIR and visible + MIR, evaluated by overall accuracy (OA) scores, for the classification of ecomorphological features. Their results show the significant contribution of the NIR and MIR bands for mapping coastal ecogeomorphological features. They also show that adding the NIR bands to the RGB band combination significantly increases the OA scores of the classifications (by +4.99% and +6.54%, with the ML and SVM algorithms, respectively). Adding the MIR bands to the combination of these bands provides classifications with even higher OA (up to 99.1% and 98.4%), demonstrating the suitability of the MIR bands for mapping coastal ecogeomorphological features.

Gadal et al. perform an analysis of coastal changes based on the use of Landsat remote sensing data sets from the 1980s to the 2020s, combined with EU geographic databases and land use plans in the coastal areas of Kaliningrad Oblast, Lithuania, and Latvia. Territorial dynamics are modeled using the GEOBIA package with object-oriented classification and machine learning algorithms (maximum likelihood, minimum distance to means, and parallelepiped classifiers) applied to multispectral images from the Landsat 5™ and Landsat 8 OLI satellites. The results were compared with the CCI Land Cover databases from 1995 and 2015.

Zaabar et al. examine LULC changes in the Allala watershed, including the city of Ténès (Algeria), using satellite images from 1999 (Landsat image), 2009 (Landsat image), and 2020 (Sentinel-2 image). In this study, the image processing and analysis is based on object-oriented classification (OBIA) using random forest (RF) and support vector machine (SVM) machine learning algorithms. Subsequently, the LULC changes were evaluated using GIS (analysis of LULC transition matrix). Their results show that RF outperforms SVM on all three input data periods, with an overall accuracy of 95.6% obtained with the Sentinel-2 image.

Modoosoodun Nicolas et al. investigate bathymetry predictions using Pleiades multispectral satellite data from around Europa Island, which is a coral island in the Mozambique Channel. The authors explore the possibility of using very high-resolution multispectral satellite data with a neural network architecture inspired by U-Net deep learning to infer bathymetry estimates. A 1 m resolution airborne LiDAR data set is used as ground truth to train the model. The predicted bathymetry values could potentially be incorporated into electronic navigational charts. Accurate results are obtained for estimating bathymetry in shallow waters, with the average absolute error not exceeding 1.5 m in this case. The authors show that this U-Net inspired deep learning technique is capable of predicting water depth from very high-resolution satellite data.

Boussetta et al. perform an assessment of the vulnerability distribution of the NE-SE coast of Jerba (Tunisia) and highlight the areas that are likely to be most affected by future sea-level rise by analyzing geophysical and socioeconomic variables processed by remote sensing approaches. They integrate remote sensing and GIS methods through supervised classification using an object-oriented approach combined with the random forest machine-learning algorithm (OBIA-RF). Their results provide baseline data on coastal zone physical processes and land use. They have found that 63% of the coastline of the northeast coast of the island has a high to very high degree of vulnerability. On the other hand, 37% of the coastline of the southeast of the island has a low to medium vulnerability.

Lam et al. propose an approach using random forest methodology on sentinel images from between 2019 and 2022 to assess the changes in the forest, planted to prevent saltwater intrusion and coastal erosion, outside the dike surrounding the coast of the Vietnamese Mekong Delta. The authors apply a method to overcome the shortcomings of optical data in identifying areas of forest loss and forest restoration. First, the satellite image data was converted into two-dimensional row and column matrix data. Each row is an object to be analyzed (pixel, object), and each column is the value of features (spectral value, vegetation value, backscatter value, observation data, etc.). Through pre-processing, optical and radar data were integrated into one dataset per year, leading to four datasets corresponding to the four observation years of 2019 to 2022. Each dataset has nine columns corresponding to three visible color bands, one near infrared band, altitude data, two polarization (VV and VH), the vegetation index (NDVI), and the observation column (class). The random forest method was found to be effective in identifying forested and non-forested areas with accuracies greater than 90%.

Rabehi et al. propose the spatial monitoring of the dynamics of Palm Beach in Algeria using the Algerian Alsat2 satellite archive (high spatial resolution (2.5 m)). They performed an evaluation "before," "during," and "after" coastal protective measures, using several spectral detection algorithms of the coastline (supervised spectral angle matching index and morphological mean filter, and unsupervised using principal/independent component analysis and NDWI normalized difference water index). The quality of this shoreline evaluation is compared to a DGPS reference shoreline in order to obtain the overall accuracy. Monitoring enables the quantitative observation of the behavior of the beach and thus allows the qualification of the degrees of effectiveness of these breakwaters. However, the performed monitoring does not take into account exceptional erosion episodes, such as storms or seasonal erosion.

Ngan et al. investigate the possibilities of assessing the current state of the shoreline and the effects of erosion on the shoreline in Vinh Chau City – Soc Trang Province, Vietnam. Using Landsat images from the years 2005 to 2022, they compare the shorelines obtained to the ones seen on Google™ Earth Pro. Google™ Earth images from the years 2006 to 2020 are then used to digitize the specific beach part of the Vinh Chau shoreline. The shoreline positions were used to determine the variation in the location of the shoreline and calculate possible erosion and accretion zones and rates of the spatial extent of the sandy beach using the digital shoreline analysis system (DSAS). The results are based on wave measurements over a short period of time, so it is not possible to come to a conclusion on the wave reduction effectiveness of current measures over the long term. Continuous monitoring during different wind seasons is needed to have a more accurate assessment of the effectiveness of wave reduction.

Bengoufa et al. propose an efficient method for the detection and extraction of coastal indicators. They use a satellite image classification approach, based on a deep learning algorithm, optimized and adapted to the extraction of a hydrodynamic and a biological coastline indicator. They use a very high-resolution Pleiades image, cut into sub-regions, and analyzed by a convolution kernel of size 3*3. Following the image processing and analysis by classification, they obtained a very high global accuracy of 92%. These results were validated by comparing them to the field surveys (reference) acquired on the same day as the satellite image acquisition. With a run-up (horizontal wave excursion) of 0.6 m, the confidence interval for the deep learning method was estimated at ±0.42 m.

In the Category of Short Articles, There Are Six Articles

James et al. propose to evaluate the evolution of a polder (Beaussais' polder located at the bottom of Beaussais' Bay on the French Emerald Coast) in the context of coastal management by drone. Using multi-spectral spectral bands (RGB, red-edge, RE, and near infrared, NIR) at very high resolution, they examine the importance of landscape cover variations by calculating several indicators, the normalized difference vegetation index (NDVI), the normalized difference water index (NDWI), and the soil adjusted vegetation index (SAVI).

Boussetta et al. study the possibility of improving the monitoring of sandy coasts by satellite observation (Ten Landsat images and two Sentinel-2 images) by applying CNN (convolutional neural network) and digital shoreline assessment system (DSAS) algorithms.

Caline et al. propose a coastal characterization of complex but representative morpho-sedimentary patterns of the seabed in Saint Malo Bay using very high-resolution topo-bathymetric LiDAR data, their morphometric derivatives, sonar, and sediment samples.

Bulot et al. present a method for mapping using a Shannon index of benthic geo-biodiversity as a whole, using continuous morphometric predictors derived from a combination of VHR topo-bathymetric LiDAR data, discrete underwater imagery responses, and 10-neural network prediction. The study took place in Saint-Malo Bay, along the coast from Saint-Briac-sur-Mer to La Pointe du Grouin.

Ternon et al. investigate the effect of local (~100 m^2) habitat components such as habitat architecture, substrate composition, and benthic community, as well as the value of photogrammetry versus visual observation on the fish community. The study was conducted in Saint-Malo Bay (Brittany, France).

Collin et al. propose an approach to spatially assess and augment the potential of natural coastal defenses using a combination of very high resolution terrain classification and network modeling. The bathymetric terrain, derived from a state-of-the-art LiDAR survey, was enhanced by a topographic position index (TPI) based classification. The study took place in Saint-Malo's lagoon, core of the French Emerald Coast, located on the Brittany coast of the English Channel, west of Mont-Saint-Michel Bay and south of Jersey Island.

Plouzané-Brest, France Simona Niculescu

References

1. Neumann B, Vafeidis AT, Zimmermann J, Nicholis RJ (2015) Future coastal population growth and exposure to sea-level rise and coastal flooding – a global assessment. PLoS ONE 10:e0118571. doi: https://doi.org/10.1371/journal.pone.0118571
2. Taherkhani M, Vitousek S, Barnard PL, Frazer N, Anderson TR, Fletcher CH (2020) Sea-level rise exponentially increases coastal flood frequency. Sci Rep 10: 6466
3. Turpie KR, Ackleson S, Bell T, Dierssen HM, Goodman J, Green R et al (2017) Spectroscopy for global observation of coastal and inland aquatic habitats. In: 2017 IEEE International Geoscience and Remote Sensing Symposium (IGARSS) (Fort Worth, TX). pp 3051–3054. doi: https://doi.org/10.1109/IGARSS.2017.8127642
4. Vousdoukas MI, Ranasinghe R, Mentaschi L, Plomaritis TA, Athanasiou P, Luijendijk A, Feyen L (2020) Sandy coastlines under threat of erosion. Nat Clim Chang 10:260–263

Contents

Detection of Coccolithophore Bloom Episodes in Algiers Bay Using Satellite and In Situ Analysis 1
Romaissa Harid, Hervé Demarcq, Shara Amanouche, Malik Ait-Kaci, Nour-El-Islam Bachari, and Fouzia Houma

Multiscale Spatiotemporal NDVI Mapping of Salt Marshes Using Sentinel-2, Dove, and UAV Imagery in the Bay of Mont-Saint-Michel, France 17
Antoine Collin, Dorothée James, Antoine Mury, Mathilde Letard, Thomas Houet, Hélène Gloria, and Eric Feunteun

Contribution of Near- and Mid-Infrared Wavebands to Mapping Fine-Scale Coastal Ecogeomorphological Features 39
Antoine Mury, Antoine Collin, Dorothée James, and Mathilde Letard

Monitoring Land Cover Change in the Southeastern Baltic Sea Since the 1980s by Remote Sensing 59
Sébastien Gadal and Thomas Gloaguen

Assessment of Land Cover Changes in the Allala Watershed Based on Object Based Image Analysis Using Landsat and Sentinel-2 Images 81
Narimane Zaabar, Simona Niculescu, and Mustapha Kamel Mihoubi

Deep Learning–Based Bathymetry Mapping from Multispectral Satellite Data Around Europa Island 97
Khishma Modoosoodun Nicolas, Lucas Drumetz, Sébastien Lefèvre, Dirk Tiede, Touria Bajjouk, and Jean-Christophe Burnel

Assessment of Coastal Vulnerability to Erosion Risk Using Geospatial and Remote Sensing Methods (Case of Jerba Island, Tunisia) 113
Amina Boussetta, Simona Niculescu, Soumia Bengoufa, Hajer Mejri, and Mohamed Faouzi Zagrarni

A Random Forest Approach for Evaluating Forest Cover Changes Outside Dikes with Sentinel Images 133
Nguyen Chi Lam, Hiep Xuan Huynh, Simona Niculescu, Quynh Do Nguyen, and Ngan Chau Vo Nguyen

Spatial Monitoring of Coastal Protection DikesCase Study of the Touristic Beach "Palm Beach, West Algiers, Algeria" 149
Walid Rabehi, Otmani Housseyn, Mohamed Amine Bouhlala, Sarah Kreri, Oussama Benabbou, Mohammed El Amin Larabi, and Hadjer Dellani

Monitoring Shoreline Changes in the Vietnamese Mekong Delta Coastal Zone Using Satellite Images and Wave Reduction Structures ... 171
Tran Van Ty, Dinh Van Duy, Huynh Thi Cam Hong, Nguyen Dinh Giang Nam, Huynh Vuong Thu Minh, Lam Van Thinh, Nguyen Vo Chau Ngan, and Nguyen Hieu Trung

Automatic Detection of Hydrodynamical and Biological Indicators of the Shoreline Using a Convolutional Neural Network 191
Soumia Bengoufa, Simona Niculescu, Mustapha Kamel Mihoubi, Rabah Belkessa, and Katia Abbad

Very High-Resolution Monitoring and Evaluation of Tidal and Ecological Restoration in Beaussais' Bay 207
Dorothée James, Antoine Collin, Antoine Mury, Mathilde Letard, O. Legal, and Alysson Lequilleuc

Assessment of Shoreline Change of Jerba Island Based on Remote Sensing Data and GIS Using DSAS Tools 213
Amina Boussetta, Simona Niculescu, Soumia Bengoufa, and Mohamed Faouzi Zagrarni

New Insights into the Shallow Morpho-Sedimentary Patterns Using High-Resolution Topo-Bathymetric Lidar: The Case Study of the Bay of Saint-Malo 219
Bruno Caline, Antoine Collin, Yves Pastol, Mathilde Letard, and Eric Feunteun

Spatial Modeling of the Benthic Biodiversity Using Topo-Bathymetric Lidar and Neural Networks 223
Angéline Bulot, Antoine Collin, Mathilde Letard, Eric Feunteun, Loic Le Goff, Yves Pastol, and Bruno Caline

Local Circalittoral Rocky Seascape Structuring Fish Community: Insights from a Photogrammetric Approach 229
Quentin Ternon, Antoine Collin, Eric Feunteun, Frédéric Ysnel, Valentin Danet, Manon Guillaume, and Pierre Thiriet

Increasing the Nature-Based Coastal Protection Using Bathymetric Lidar, Terrain Classification, Network Modelling: Reefs of Saint-Malo's Lagoon? 235
Antoine Collin, Yves Pastol, Mathilde Letard, Loic Le Goff, Julien Guillaudeau, Dorothée James, and Eric Feunteun

Abstracts of Keyspeakers 243

Index 247

Contributors

Katia Abbad National Higher Institute of Marine and Coastal Sciences, Algiers, Algeria

Malik Ait-Kaci ECOSYSMarL: Laboratoire des Écosystèmes Marins et Littoraux, École Nationale Supérieure des Sciences de la Mer et de l'Aménagement du Littoral (ENSSMAL), Campus Universitaire de Dely Ibrahim Bois des Cars, Alger, Algeria

Shara Amanouche APPL: Agence de la Protection et de la Promotion du Littoral Algérois, Alger, Algeria

Nour-El-Islam Bachari USTHB: Université des Sciences et Technologie Houari Boumedien, Département Écologie et Environnement, Alger, Algeria

Touria Bajjouk Ifremer, Dynamiques des Ecosystèmes Côtiers, LEBCO, Plouzané, France

Rabah Belkessa National Higher Institute of Marine and Coastal Sciences, Algiers, Algeria

Oussama Benabbou Centre des Techniques Spatiales, Algerian Space Agency, Arzew, Algeria

Soumia Bengoufa University of Western Brittany, CNRS, LETG Brest UMR 6554 CNRS, Technopôle Brest-Iroise, Av. Dumont d' Urville, Plouzané-Brest, France

National School of Marine Sciences and Coastal Management (ENSSMAL), University Campus of Dely Ibrahim Bois des Cars, Algiers, Algeria

Mohamed Amine Bouhlala Centre des Techniques Spatiales, Algerian Space Agency, Arzew, Algeria

Amina Boussetta U.R Applied Hydrosciences UR13ES81, Higher Institute of Water Sciences and Techniques, University of Gabes, Cité Erryadh, Gabes, Tunisia

Angéline Bulot Ecole Pratique des Hautes Etudes – PSL Université, LETG-Dinard UMR 6554, Dinard, France

Jean-Christophe Burnel Université Bretagne Sud, UMR CNRS IRISA, Vannes, France

Bruno Caline Sedimentologist, Dinard, France

Antoine Collin Ecole Pratique des Hautes Etudes – PSL University, Coastal GeoEcological Lab, Dinard, France

Ecole Pratique des Hautes Etudes – PSL Université, Dinard, France

Laboratoire d'excellence CORAIL, Perpignan, France

LabEx CORAIL, Papetoai, French Polynesia

Ecole Pratique des Hautes Etudes – PSL Université, LETG-Dinard UMR 6554, Dinard, France

Valentin Danet Museum National d'Histoire Naturelle (MNHN), Station Marine de Dinard, Centre de Recherche et d'Enseignement sur les Systèmes CÔtiers (CRESCO), Dinard, France

Hadjer Dellani Centre des Techniques Spatiales, Algerian Space Agency, Arzew, Algeria

Hervé Demarcq MARBEC, IRD, Ifremer, CNRS, Univ Montpellier, Sète, Sète Cedex, France

Quynh Do Nguyen Can Tho University, Can Tho City, Vietnam

Lucas Drumetz IMT Atlantique, UMR CNRS Lab-STICC, Brest, France

Eric Feunteun Museum National d'Histoire Naturelle, Ecole Pratique des Hautes Etudes, CNRS BOREA, Dinard, France

Museum National d'Histoire Naturelle, CNRS BOREA, Dinard, France

Museum National d'Histoire Naturelle (MNHN), Station Marine de Dinard, Centre de Recherche et d'Enseignement sur les Systèmes CÔtiers (CRESCO), Dinard, France

Laboratoire Biologie des Organismes et Écosystèmes Aquatiques (UMR BOREA), MNHN, CNRS, Sorbonne Université, Université de Caen Normandie, Université des Antilles, IRD, Paris Cedex, France

Ecole Pratique des Hautes Etudes – PSL University, Coastal GeoEcological Lab, Dinard, France

Sébastien Gadal CNRS, ESPACE, UMR 7300, Aix-Marseille Université, Université Côte-d'Azur, Avignon Université, Avignon, France

North-Eastern Federal University, Republic of Sakha, Yakutia, Yakutsk, Russian Federation

Thomas Gloaguen CNRS, ESPACE, UMR 7300, Aix-Marseille Université, Université Côte-d'Azur, Avignon Université, Avignon, France

Kaunas University of Technology, Faculty of Civil Engineering and Architecture, Cultural and Spatial Environment Research Group, Kaunas, Lithuania

Hélène Gloria Ecole Pratique des Hautes Etudes – PSL University, Coastal GeoEcological Lab, Dinard, France

Julien Guillaudeau Museum National d'Histoire Naturelle, CNRS BOREA, Dinard, France

Manon Guillaume Museum National d'Histoire Naturelle (MNHN), Station Marine de Dinard, Centre de Recherche et d'Enseignement sur les Systèmes CÔtiers (CRESCO), Dinard, France

Romaissa Harid ECOSYSMarL: Laboratoire des Écosystèmes Marins et Littoraux, École Nationale Supérieure des Sciences de la Mer et de l'Aménagement du Littoral (ENSSMAL), Campus Universitaire de Dely Ibrahim Bois des Cars, Alger, Algeria

Huynh Thi Cam Hong Department of Hydraulic Engineering, Can Tho University, Can Tho City, Vietnam

Thomas Houet Université of Rennes 2, CNRS LETG, Rennes, France

Fouzia Houma ECOSYSMarL: Laboratoire des Écosystèmes Marins et Littoraux, École Nationale Supérieure des Sciences de la Mer et de l'Aménagement du Littoral (ENSSMAL), Campus Universitaire de Dely Ibrahim Bois des Cars, Alger, Algeria

Otmani Housseyn National School of Marine Sciences and Coastal Management – Campus Universitaire de Dely Ibrahim, Algiers, Algeria

Hiep Xuan Huynh Can Tho University, Can Tho City, Vietnam

Dorothée James Ecole Pratique des Hautes Etudes – PSL University, Coastal GeoEcological Lab, Dinard, France

Sarah Kreri Centre des Techniques Spatiales, Algerian Space Agency, Arzew, Algeria

Nguyen Chi Lam Can Tho University, Can Tho City, Vietnam

Mohammed El Amin Larabi Centre des Techniques Spatiales, Algerian Space Agency, Arzew, Algeria

Sébastien Lefèvre Université Bretagne Sud, UMR CNRS IRISA, Vannes, France

Loic Le Goff Museum National d'Histoire Naturelle, CNRS BOREA, Dinard, France

O. Legal Ecole Pratique des Hautes Etudes – PSL University, Coastal GeoEcological Lab, Dinard, France

Alysson Lequilleuc Ecole Pratique des Hautes Etudes – PSL University, Coastal GeoEcological Lab, Dinard, France

Mathilde Letard Ecole Pratique des Hautes Etudes – PSL University, Coastal GeoEcological Lab, Dinard, France

Ecole Pratique des Hautes Etudes – PSL Université, Dinard, France

Ecole Pratique des Hautes Etudes – PSL Université, LETG-Dinard UMR 6554, Dinard, France

Hajer Mejri U.R Applied Hydrosciences UR13ES81, National Engineering School of Gafsa (ENIGA), University of Gafsa, Gafsa, Tunisia

Mustapha Kamel Mihoubi ENSH, National Higher School of Hydraulics, MVRE, Laboratoire de Mobilisation et valorisation des ressources en eau, Blida, Algeria

National Higher Institute of Hydraulics, Blida, Algeria

Huynh Vuong Thu Minh Department of Water Resources, Can Tho University, Can Tho City, Vietnam

Antoine Mury Ecole Pratique des Hautes Etudes – PSL University, Coastal GeoEcological Lab, Dinard, France

Ecole Pratique des Hautes Etudes – PSL Université, Dinard, France

UMR AMAP, IRD, CIRAD, CNRS, INRAE, Université de Montpellier, Montpellier, France

Nguyen Dinh Giang Nam Department of Water Resources, Can Tho University, Can Tho City, Vietnam

Nguyen Vo Chau Ngan Department of Water Resources, Can Tho University, Can Tho City, Vietnam

Ngan Chau Vo Nguyen Can Tho University, Can Tho City, Vietnam

Khishma Modoosoodun Nicolas Université Bretagne Sud, UMR CNRS IRISA, Vannes, France

Paris Lodron University of Salzburg, Salzburg, Austria

Simona Niculescu Laboratory LETG-Brest, UMR 6554 CNRS, University of Western Brittany, Plouzané, France

Yves Pastol Service Hydrographique et Océanographique de la Marine, Brest, France

Walid Rabehi Centre des Techniques Spatiales, Algerian Space Agency, Arzew, Algeria

Quentin Ternon Museum National d'Histoire Naturelle (MNHN), Station Marine de Dinard, Centre de Recherche et d'Enseignement sur les Systèmes CÔtiers (CRESCO), Dinard, France

Laboratoire Biologie des Organismes et Écosystèmes Aquatiques (UMR BOREA), MNHN, CNRS, Sorbonne Université, Université de Caen Normandie, Université des Antilles, IRD, Paris Cedex, France

Pierre Thiriet Patrimoine Naturel (UAR OFB–CNRS–MNHN), MNHN – Station Marine de Dinard, Dinard, France

Dirk Tiede Paris Lodron University of Salzburg, Salzburg, Austria

Nguyen Hieu Trung Institute of Climate Change, Can Tho University, Can Tho City, Vietnam

Dinh Van Duy Department of Hydraulic Engineering, Can Tho University, Can Tho City, Vietnam

Lam Van Thinh Department of Water Resources, Can Tho University, Can Tho City, Vietnam

Tran Van Ty Department of Hydraulic Engineering, Can Tho University, Can Tho City, Vietnam

Frédéric Ysnel Université de Rennes 1, UMR BOREA, MNHN, Rennes Cedex, France

Narimane Zaabar University of Western Brittany, CNRS, LETG Brest UMR 6554 CNRS, Technopôle Brest-Iroise, Plouzané-Brest, France

ENSH, National Higher School of Hydraulics, MVRE, Laboratoire de Mobilisation et valorisation des ressources en eau, Blida, Algeria

Mohamed Faouzi Zagrarni U.R Applied Hydrosciences UR13ES81, Higher Institute of Water Sciences and Techniques, University of Gabes, Cité Erryadh, Gabes, Tunisia

Detection of Coccolithophore Bloom Episodes in Algiers Bay Using Satellite and In Situ Analysis

Romaissa Harid, Hervé Demarcq, Shara Amanouche, Malik Ait-Kaci, Nour-El-Islam Bachari, and Fouzia Houma

Abstract In Algiers Bay, coccolithophore blooms of *Holococcolithophora sphaeroidea* species were identified from in situ observations during August 2003, July–August 2013, July 2015, and July 2017. This study determines for the first time in Algiers Bay the episodes of coccolithophore blooms from 2003 to 2018 using satellite and in situ observations. In addition, a new coccolithophore remote sensing reflectance index (Cocco-Index) is presented, which aims to detect the presence of coccolithophore bloom from satellites in space and time. It was applied to 16 years of data from the Moderate Resolution Imaging Sensor (2003–2018). From 2003 to 2018, the coccolithophore bloom appeared yearly in Algiers Bay but with a remarkable seasonal variability, developing mainly in winter and rarely in summer. This work is the first demonstration of applying a coccolithophore index for this region over such a large timescale

Keywords Coccolithophores index · Remote sensing reflectance · MODIS · Spectral response · Algiers Bay

R. Harid (✉) · M. Ait-Kaci · F. Houma
ECOSYSMarL: Laboratoire des Écosystèmes Marins et Littoraux, École Nationale Supérieure des Sciences de la Mer et de l'Aménagement du Littoral (ENSSMAL), Campus Universitaire de Dely Ibrahim Bois des Cars, Alger, Algeria
e-mail: r.harid@enssmal.dz; m.ait-kaci@enssmal.dz; f.houmabachari@enssmal.dz

H. Demarcq
MARBEC, IRD, Ifremer, CNRS, Univ Montpellier, Sète, Sète Cedex, France
e-mail: herve.demarcq@ird.fr

S. Amanouche
APPL: Agence de la Protection et de la Promotion du Littoral Algérois, Alger, Algeria
e-mail: amenouche.s@appl-dz.com

Nour-El-Islam Bachari
USTHB: Université des Sciences et Technologie Houari Boumedien, Département Écologie et Environnement, Alger, Algeria
e-mail: n.bechari@usthb.dz

S. Niculescu (ed.), *European Spatial Data for Coastal and Marine Remote Sensing*, https://doi.org/10.1007/978-3-031-16213-8_1

1 Introduction

Coccolithophores species dominate primary production in oligotrophic areas [37, 42], trapping CO_2 and modifying the carbon cycle [8]. Additionally, the coccolithophore may result in artifacts into ^{14}C-derived physiological values [28]. The shells of these species are deposited on the seafloor and eventually become part of the marine fossil archive. Coccolithophores are interesting microorganisms that can be used to understand the causes of changes in the global climate. Coccolithophores largely appear in the marine environment. Generally, they have two equal or subequal smooth flagella and an extracellular envelope with one or more layers of organic shells associated with inorganic compounds located outside the organism (Fig. 1).

The shape and structure of coccolithophores play several roles in the marine ecosystem, such as marine biochemical reactions. Calcification produces bicarbonate (HCO_3) used in the photosynthesis process. Moreover, these species regulate the light field. They reflect ultraviolet rays away from the cell, protecting them from photodamage and living near the surface waters. In addition, depending on their shape, they could refract the light toward the inside of the cell, which allows them to harvest light in deep water. The coccolithophores species represent a very high diversity in the hot oligotrophic areas [44]. They represent around 50% of the oceanic carbon pump in the offshore domain [40] and are a major producer of particulate inorganic carbon (PIC) in the pelagic zone of oceanic regimes [4, 25].

In ocean color remote sensing, the spectral response of limestone species (such as coccolithophores) is characterized by a high back-scattering capacity caused by their limestone shells. The identification of coccolithophore species during bloom episodes requires signal discrimination from the back-scattering properties of optically significant particles [1, 2, 5, 6, 17–19, 24, 35]. Several satellite sensors are dedicated to measuring the ocean color signal, allowing daily observation of marine

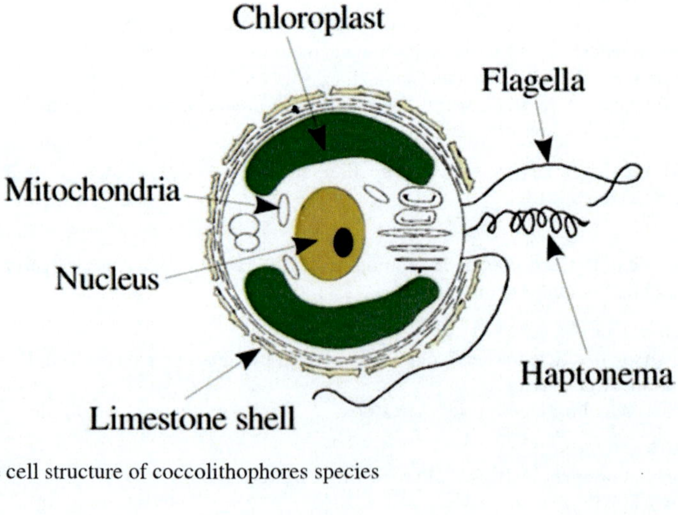

Fig. 1 The cell structure of coccolithophores species

surfaces globally. Various techniques have been used to detect coccolithophores from space, such as spectral band ratio algorithms, indexes based on remote sensing reflectance (Rrs) ratios [3], and coccolithophore pixel classification [35]. Moore et al. [30] extended an optical water-type classification scheme (developed in [29]) by adding a new coccolithophore bloom class created from extracted satellite Rrs. The remote sensing reflectance (Rrs) is the basic product of the ocean color sensors [21]. The Rrs defines the water-leaving radiance normalized by the sun in the vertical without atmosphere, relative to the incident irradiance of the sun at the zenith.

Several studies have used suspended particulate matter (SPM) as the most suitable element to quantify when detecting coccolithophore blooms. Perrot et al. [36] identified coccolithophore blooms from space using a spectral method to discriminate the SPM fraction corresponding to coccoliths from a total non-algal SPM signal. Moore et al. [30] used a key difference between the coccolithophore bloom and SPM: the peak wavelength for the reflected light, which is 555 nm for SPM and 490 nm for coccolithophore bloom. In the same wavelength range, Balch et al. [4] included the back-scattering due to coccolithophores at 443 and 555 nm to quantify the PIC concentration from space in the same wavelength range.

Coccolithophore blooms have specific bio-optical properties, which helps distinguish them from spaceborne sensors. Remote sensing reflectance classification techniques are often used to detect the presence of coccolithophore blooms [1]. Most phytoplankton species have a common spectral response curve and cannot be identified individually from the space [23]. Some species, however, have unique spectral response curves that enable their identification. Coccolithophores are a special class of phytoplankton. They grow and deposit calcium carbonate shells outside their cells when they bloom. Coccolithophores can impact the sensing signal used to estimate the chlorophyll-*a* concentrations by standard empirical algorithms [11, 16, 22].

To study the coccolithophore bloom episodes in Algiers Bay, we used an approach based on two complementary methods: in situ and satellite observations. The first aim of this work is to study the coccolithophore bloom episodes observed several times in Algiers Bay based on in situ observations. Few studies on coccolithophore species in Algiers Bay have been published to date. However, other data sources, such as satellite observations, can be used to monitor the flowering of coccolithophores. The in situ observations were monitored in space and time from satellite data. A coccolithophore index based on remote sensing reflectance was also established to define for the first time in Algiers Bay the coccolithophore bloom pixel in the corresponding observations between 2003 and 2018.

2 Study Area

Algiers Bay is located in the central part of the Algerian coast (Africa). It extends over 70 km of the coastline, between Caxine Cape in the west, Matifou Cape in the east, the Mediterranean Sea in the north, and Algiers City in the south (Fig. 2). Two

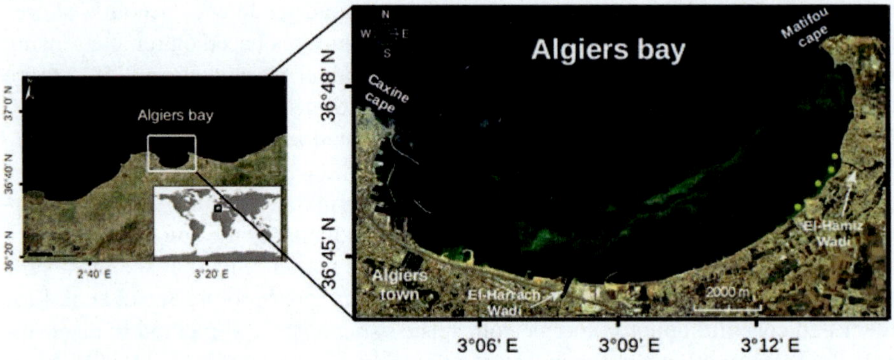

Fig. 2 Sentinel-2 map presentation of the Algiers Bay (the points in yellow are the in situ sampled stations on July 19, 2017)

main wadis flow into the bay, the El-Harrach wadi and the El-Hamiz wadi, whose flow is very low [15]. Some coastal protection agencies monitor potentially dangerous microalgae in Algiers Bay, such as *Agence de la Protection et de la Promotion du Littoral Algérois* (APPL).

3 Materials and Methods

3.1 In Situ Data

We collected water samples (4 in situ stations (Fig. 2)) on the day of bloom onset and preserved them in a Lugol's solution to determine the species responsible for the bay coloration on July 19, 2017. In the laboratory, the samples were sedimented in the special phytoplankton tubes. They were then analyzed under an inverted microscope of type ZEIZZ using the NF EN 15204 method (Utermöhl method) [33], under an objective of X40. Figure 2 shows the coccolithophore bloom during July 2017 from Sentinel-2 MSI data.

3.2 Satellite Data Acquisition

We used the MODerate resolution Imaging Sensor (MODIS)/AQUA Level-2 data from NASA's Ocean Color Web [32] to study the coccolithophore bloom episodes in the Algiers Bay between 2003 and 2018. The MODIS makes daily observations in Algiers Bay. This study used the MODIS Rrs in the visible wavelengths (412, 443, 469, 488, 531, 547, 555, 645, 667, 678 nm) to define a coccolithophore bloom index. The data set consists of 5844 daily observations through 15,020 individual orbits of Rrs MODIS Level-2 data. All data were re-projected in a rectangular grid of 1 km between 36.2°N 38°N and 2.1°E 4.1°E.

3.3 Coccolithophore Bloom Index (Cocco-Index)

The ocean color remote sensing provides large spatial and temporal observations in the Algiers Bay, enabling us to detect from space the episodes of the coccolithophore bloom in the time series. Therefore, we used the optical proprieties to develop an index to detect the episodes of coccolithophore bloom from satellite observations between 2003 and 2018 in Algiers Bay. To our knowledge, the coccolithophore remote sensing reflectance index for Algiers Bay proposed in this work is the first presented in the literature. It aims to detect the coccolithophore bloom from the daily MODIS data.

The Cocco-Index proposed in this paper is based on a threshold of Rrs between the green (531 nm, 547 nm, and 555 nm) and the blue (443 nm, 469 nm, and 488 nm) bands for the MODIS. These thresholds (Eq. 1) are fixed according to the maximum values of Rrs, which increase uniformly in the blue and green bands [13].

The Cocco-Index determines the position of bloom pixels (BPs) by calculating the intersection between the Rrs ≥ 0.008 in the blue band and the Rrs ≥ 0.004 in the green band, as shown in the following formula:

$$\text{Cocco}-\text{Index}=\int_{\text{Rrs}\geq 0.008}^{\text{Rrs}^{\max}}\text{Rrs}(\text{blue})\cap\int_{\text{Rrs}\geq 0.004}^{\text{Rrs}^{\max}}\text{Rrs}(\text{green}) \tag{1}$$

After several tests, the thresholds 0.004 (in green) and 0.008 (in blue) were fixed. There are no quality flags applied to the reflectance data. Figure 3 shows that the SPM and coccolithophore spectral response difference is marked with Rrs values and bands. This index aims to determine the BPs of coccolithophore from the standard Rrs MODIS Level-2 data (so that everyone can use it, even beginners in remote sensing).

4 Results

4.1 Coccolithophore Bloom Episodes in the Algiers Bay from In Situ Observations

Our analysis's first quantitative and qualitative results reveal that the yellowish coloration in Algiers surface water during July 19, 2017, was caused by the proliferation of the *Holococcolithophora sphaeroidea* species. The maximum bloom concentration is $65*10^6$ cells/liter. Higher concentrations were also observed in August 2003 in Algiers by $4.7*10^8$ cells/liter [20]. The green color shown in Fig. 2 reflects the high back-scattering of the limestone shell of the coccolithophore.

Historical *Holococcolithophora sphaeroidea* episodes were determined from the literature. However, in 2003, scientists signaled the first appearance of the *Holococcolithophora sphaeroidea* bloom (coccolithophores species) in Algiers Bay

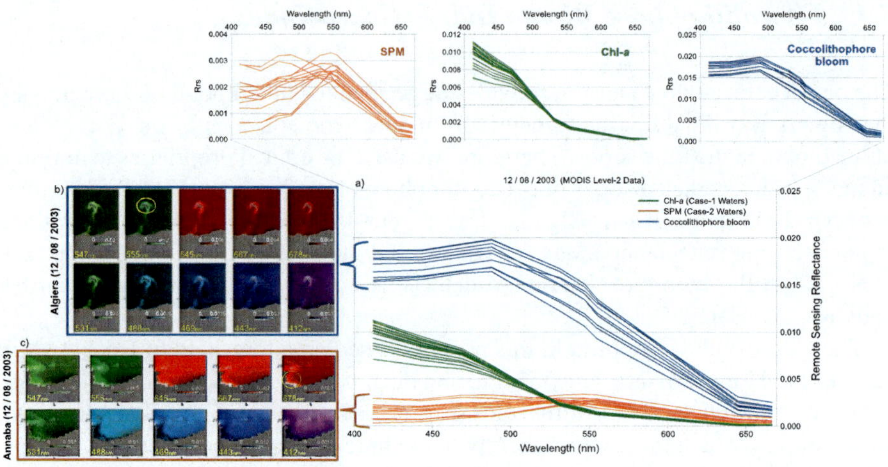

Fig. 3 Spectral signature of Chl-a, SPM, and coccolithophore bloom in the Algerian coastal waters. The Rrs used in this figure are the MODIS Level-2 data and correspond to the coccolithophore bloom day of Algiers Bay in August 12, 2003 [20]. (**a**) Rrs versus wavelengths in three different water categories: (i) Cas-1 waters in green curves, (ii) Cas-2 waters (coastal waters of Annaba Bay) in brown curves, and (iii) coccolithophore bloom waters (coastal waters of the Algiers Bay) in blue curves. (**b**) Spatial distribution of Rrs for each wavelength on the coccolithophore bloom day of Algiers Bay (August 12, 2003). (**c**) Spatial distribution of Rrs for each wavelength in Annaba Bay (August 12, 2003)

Table 1 Coccolithophore bloom episodes in Algiers Bay from in situ observations

Episode	Month	Year	Author	Bloom species
E1	August	2003	Illoul et al. (2008)	*Holococcolithophora sphaeroidea* (coccolithophore species)
E2	July and August	2013	APPL (2013)	*Holococcolithophora sphaeroidea* (coccolithophore species)
E3	July	2013	APPL (2013)	*Holococcolithophora sphaeroidea* (coccolithophore species)
E4	July	2017	Present study	*Holococcolithophora sphaeroidea* (coccolithophore species)

[20]. This bloom appeared in the Mediterranean Sea for the first time in 2001 in Spain [7] and Algeria in August 2003 [20]. The other episodes were observed during August and July 2013, July 2015 (APPL, personnel communication), and July 2017 (our work). Table 1 summarizes the episodes of the *Holococcolithophora sphaeroidea* bloom between 2003 and 2018 observed from in situ measurements in Algiers Bay.

4.2 Monitoring and Spectral Response of E1, E2, and E3

We used satellite observations to monitor the spatial and temporal extent of the episodes listed in Table 1. The MODIS data corresponding to the in situ observations (Table 1) corresponds available to 2003, 2013, and 2015 episodes (not available for July 2017). In August 2003, the bloom was detached from the coast and moved toward the offshore area (Fig. 4a–c). However, during the 2013 episode (Fig. 5), the bloom remained attached to the coast and was propagated eastward by a cyclonic eddy [41] (Fig. 5a–c). Finally, during 2015 episode (Fig. 6), the bloom remained inside the bay, while a small plume was (Fig. 6i) detached from the coast to the bay's northeast (Fig. 6a–d).

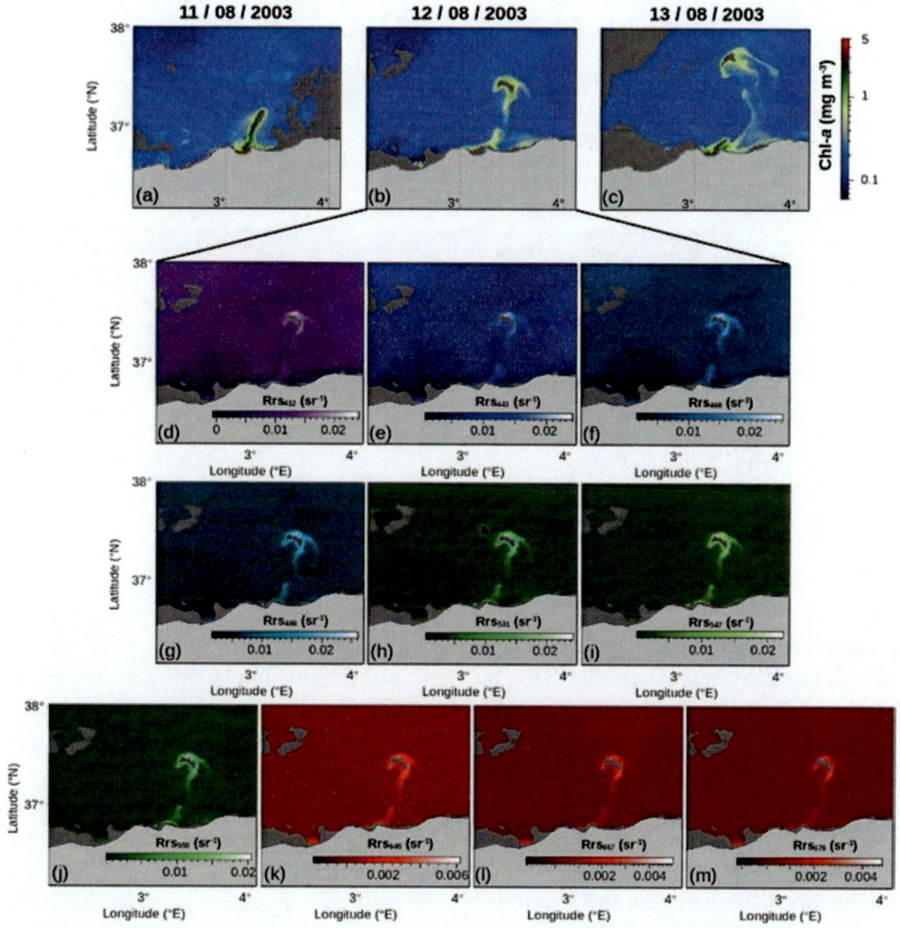

Fig. 4 Monitoring of coccolithophore bloom during August 2003 from MODIS satellite data (**a–c**). From **d–m** are the Rrs of August 12, 2003, at 412, 443, 469, 488, 531, 547, 555, 645, 667, and 687 nm wavelengths, respectively

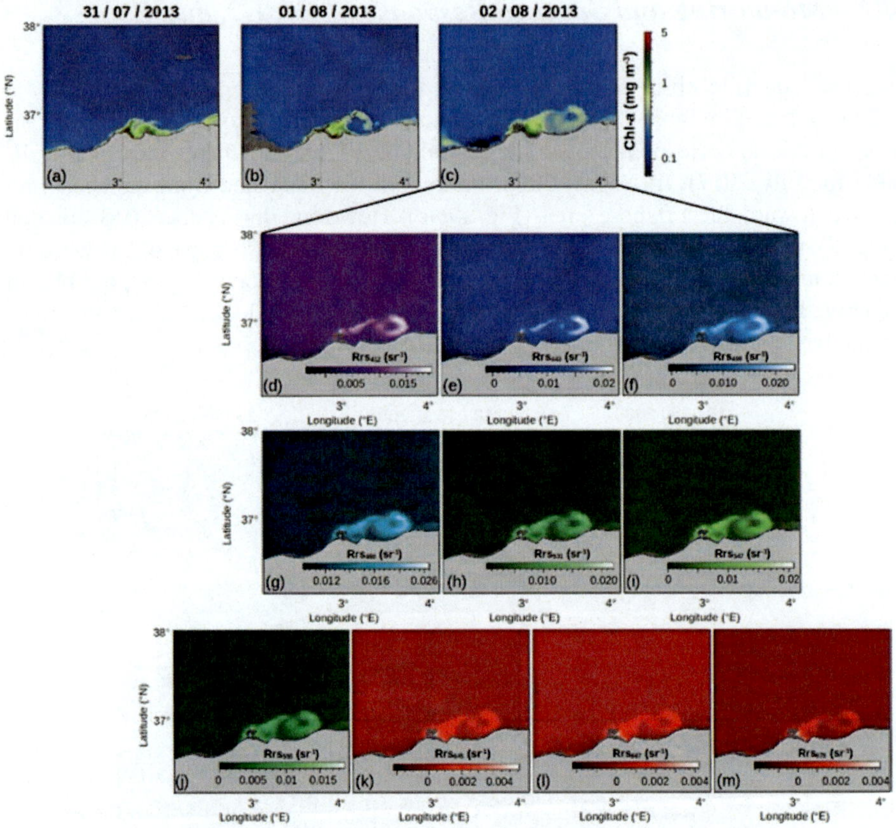

Fig. 5 Monitoring of coccolithophore bloom during July and August 2013 from MODIS satellite data (**a–c**). From **d** to **m** are the Rrs of August 02, 2013, at 412, 443, 469, 488, 531, 547, 555, 645, 667, and 687 nm wavelengths, respectively

Figures 4, 5, and 6 indicates the Rrs in the visible wavelengths during E1, E2, and E3, respectively. Elevated Rrs signal at the green and blue bands is observed for each episode (see Figs. 4, 5, and 6). The maximum Rrs (for all wavelengths) inside the Algiers Bay represents the BPs, but with a different number of BPs for each band (see the Rrs matchup in Figs. 4, 5, and 6). The early works on the optical properties of coccolithophore blooms have also shown high Rrs in the blue and green bands [5, 6, 18].

However, the number of BPs in the green bands is higher than in the blue bands (Figs. 4, 5, and 6). The maximum values of Rrs are observed at wavelengths 488 nm and 531 nm, respectively (Figs. 4g, h, 5g, h, and 6h, i). The blue curve in Fig. 3a shows this difference, but the maximum number of BPs is observed at the 555 nm band (see Figs. 4j, 5j, and 6k).

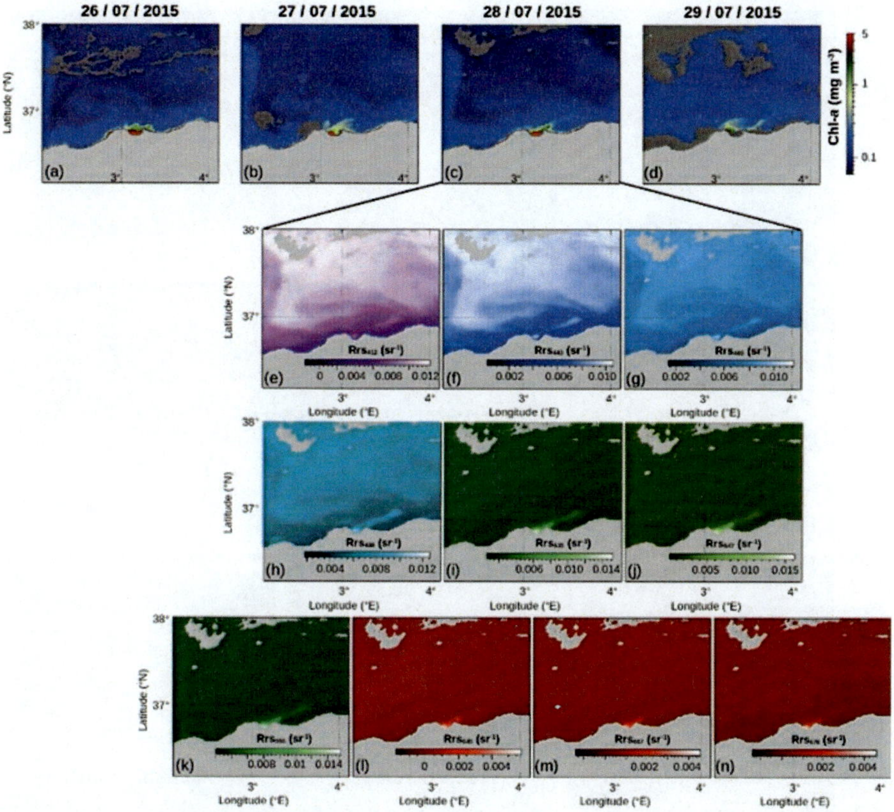

Fig. 6 Monitoring of coccolithophore bloom during July 2015 from MODIS satellite data (**a–c**). From **d** to **m** are the Rrs of July 28, 2015, at 412, 443, 469, 488, 531, 547, 555, 645, 667, and 687 nm wavelengths, respectively

4.3 *Cocco-Index of E1, E2, and E3*

The Cocco-Index was applied to the Rrs corresponding to the in situ observations of coccolithophore blooms during the first episode (E1 in Table 1) and successfully detected the BP positions of E1, E2, and E3. Figure 7 indicates the results obtained. Therefore, this index can be applied to the daily MODIS Level-2 data in the Algiers Bay surface waters.

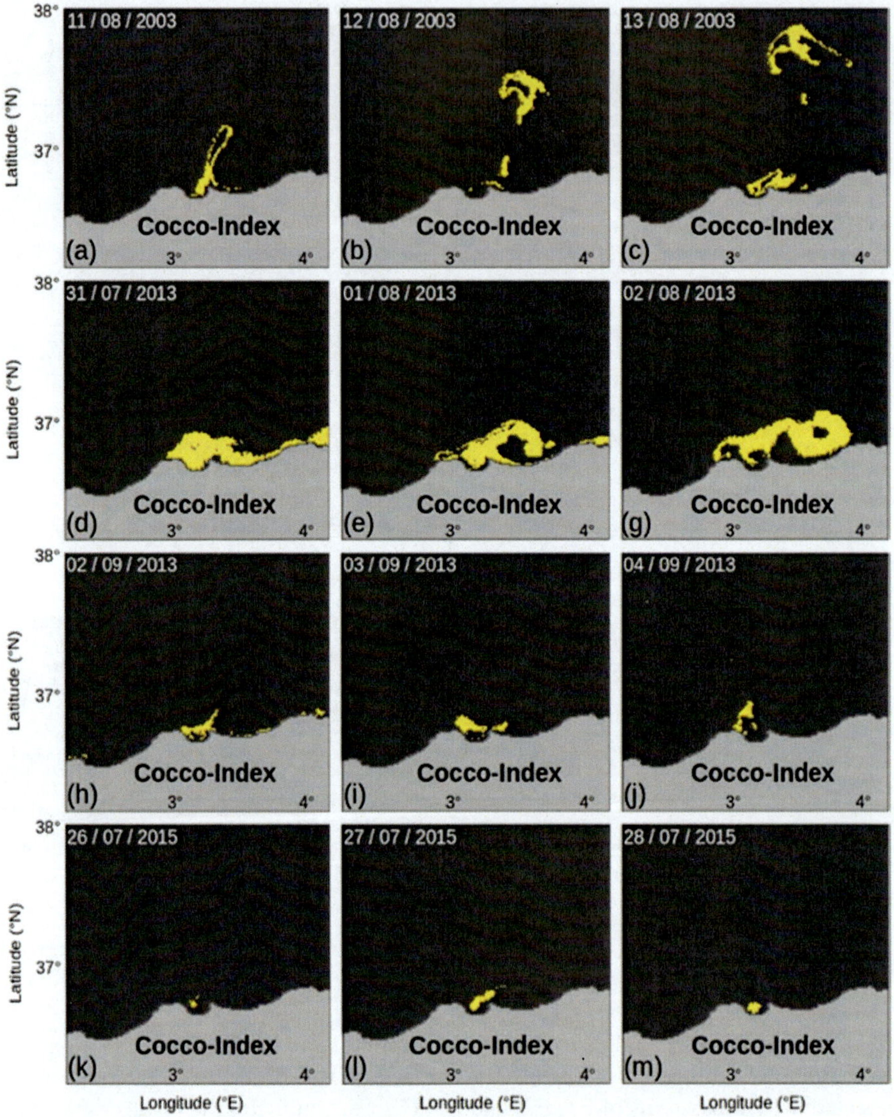

Fig. 7 Cocco-Index corresponding to the bloom episodes cited in Table 1

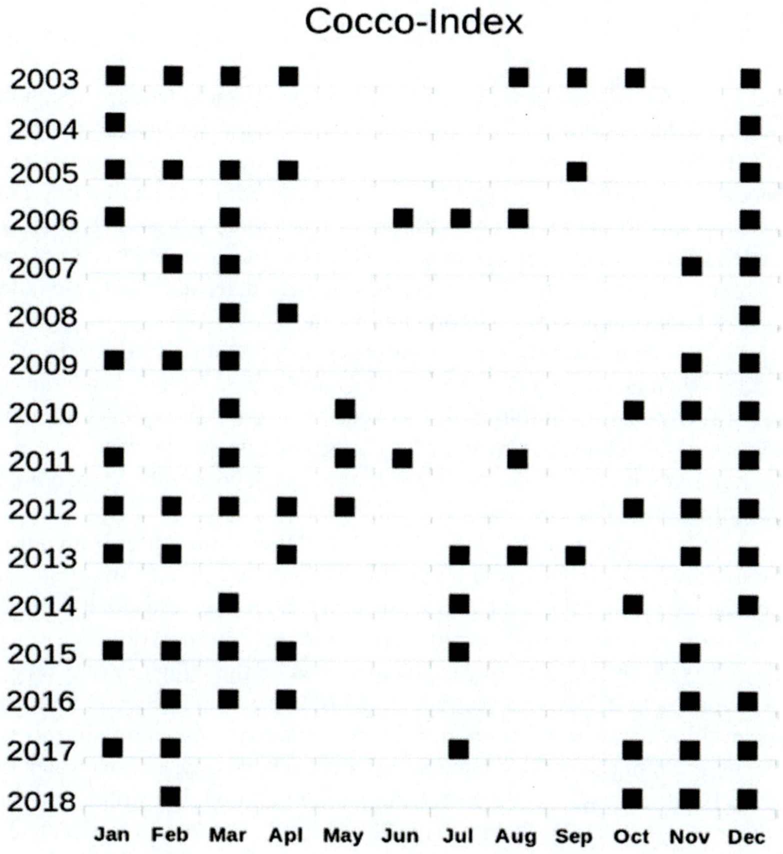

Fig. 8 Coccolithophore bloom episodes in the Algiers Bay between 2003 and 2018 from MODIS observations

4.4 Coccolithophore Bloom Episodes in the Algiers Bay (from 2003 to 2018)

From 2003 to 2018, the Cocco-Index appeared yearly in Algiers Bay but with remarkable seasonal variability (Fig. 8). Figures reflecting the days when coccolithophore blooms were detected for the episodes in Fig. 8 were illustrated in Harid [14] and can be downloaded from https://doi.org/10.5281/zenodo.6612047.

The results indicate that the coccolithophore blooms appear in winter and rarely in summer (Fig. 8). This bloom was observed during December of each year from 2003 to 2018, except in 2015 (Fig. 8). On the other hand, this bloom was observed almost every year in January and February (Fig. 8). During June, the bloom was developed only in 2006 and 2011 (Fig. 8). Generally, the coccolithophore bloom is observed for three successive months every year (Fig. 8).

5 Discussion

The limestone shell of the coccolithophore species is characterized by a high backscattering of light, resulting in the characteristic elevated Rrs in the 488 nm and 555 nm wavebands. In the red wavelengths, the Rrs is the weakest compared to the other bands [27, 31] (Figs. 4, 5, and 6). However, the BP identified by our Cocco-Index corresponds to the high concentration of particulate inorganic carbon (PIC) downloaded from NASA's Ocean Color Web [32] for each episode (data not shown in this paper). In contrast, the Cocco-Index can only determine and calculate the area of BP, which matches coccolithophore bloom.

The proliferation of coccolithophore species is more characteristic of the Algiers Bay in winter than in summer (Fig. 8). Probably, the winter mixing water caused by eddies [26] favors the proliferation of this species. Moreover, the coccolithophore class dominates the identified phytoplankton species in the Mediterranean Sea [34]. The Algerian basin represents less than 3% of the BPs of coccolithophore compared to the rest of the global oceans [30].

It is important to note that the surface waters of Algiers Bay suffer from pollution [39] due to human activity (harbor and industry) and very significant urbanization in the coastal area [38]. Figure 2 indicates that this bay is semi-closed. It traps nutrients inside its waters, favoring coccolithophore blooms' proliferation.

During the summer, Algiers Bay is characterized by the influence of a very high level of human activity. The El-Harrach and El-Hamiz wadis (Fig. 2) and certain domestic and industrial inputs flow directly into the bay, thus increasing its nutrients. The increase in sea surface temperature over several days causes water stratification [12] and can explain the coccolithophore bloom episodes observed during the summer of 2003, 2006, 2011, 2013, 2014, 2015, and 2017 (see Table 1 and Fig. 8) in Algiers Bay. However, the *Holococcolithophora sphaeroidea* tolerate the high temperatures and intense oligotrophic marine environments of the Mediterranean Sea [9, 10, 43] such as the Algerian Basin [15] and are more abundant in subsurface seawater [10].

6 Conclusion

Based on the results obtained in this study, we can now identify the occurrence of the coccolithophore bloom from ocean color remote sensing data in the surface waters of Algiers Bay. This approach is useful when we cannot collect in situ samples to identify and monitor the coccolithophore bloom. The proposed index is limited to detecting coccolithophore blooms and not other phytoplankton groups. We cannot apply this index to the entire Algerian coastal waters because it was created using the spectral response, which is only specific to the Algiers Bay surface waters.

The study of this bloom's seasonal and interannual variability in large time series of observations allowed us to better predict the future impact of climate change. In

Algiers Bay, coccolithophore bloom episodes were marked yearly (from 2003 to 2018), generally more observed in winter.

In conclusion, satellite data are a good tool for monitoring coccolithophore blooms in Algiers Bay. Bloom episodes appear in the Bay at least two or three times a year. However, this work should be followed by studies on the interactions between biological, physical, chemical, and environmental factors in the coccolithophore bloom areas.

Acknowledgments We applied the sequence-determines-credit (SDC) approach for the authors' sequence. We would like to thank the space agency NASA for providing the MODIS satellite images and the Copernicus service for providing the Sentinel data used in this paper. The authors thank Dr. Mehdia-Asma Keraghel (ENSSMAL, Algiers, Algeria), Dr. Hosseyn Otmani (ENSSMAL, Algiers, Algeria), Mr. Mohamed Zerrouki (ENSSMAL, Algiers, Algeria), and Mrs. Roufeida Benharkou (University of Annaba, Algeria) for their assistance in this work. This research was supported by the MESRS (Algerian government) PhD scholarship.

References

1. Ackleson SG, Balch WM, Holligan PM (1994) Response of water-leaving radiance to particulate calcite and chlorophyll a concentrations: a model for Gulf of Maine coccolithophore blooms. J Geophys Res 99:7483–7499. https://doi.org/10.1029/93JC02150
2. Alvain S, Moulin C, Dandonneau Y, Loisel H (2008) Seasonal distribution and succession of dominant phytoplankton groups in the global ocean: a satellite view: phytoplankton groups – a satellite view. Global Biogeochem Cycles 22. https://doi.org/10.1029/2007GB003154
3. Amin R, Gilerson A, Gross B, Moshary F, Ahmed S (2009) MODIS and MERIS detection of dinoflagellates blooms using the RBD technique. In: Bostater CR Jr, Mertikas SP, Neyt X, Velez-Reyes M (eds) Presented at the SPIE Europe remote sensing, Berlin, p 747304. https://doi.org/10.1117/12.830631
4. Balch WM, Gordon HR, Bowler BC, Drapeau DT, Booth ES (2005) Calcium carbonate measurements in the surface global ocean based on Moderate-Resolution Imaging Spectroradiometer data. J Geophys Res Oceans 110. https://doi.org/10.1029/2004JC002560
5. Brown C (1995) Global distribution of coccolithophore blooms. Oceanography 8:59
6. Brown CW, Yoder JA (1994) Coccolithophorid blooms in the global ocean. J Geophys Res Oceans 99:7467–7482. https://doi.org/10.1029/93JC02156
7. Cros L, Fortuño JM (2002) Atlas of Northwestern Mediterranean Coccolithophores. Sci Mar 66:1–182. https://doi.org/10.3989/scimar.2002.66s11
8. Cros L, Fortuño JM, Estrada M (2013) Elemental composition of coccoliths: Mg/Ca relationships. Sci Mar 77:63–67. https://doi.org/10.3989/scimar.03727.27E
9. Cros Miguel L (2001) Planktonic coccolithophores of the NW Mediterranean. Departament d'Ecologia, Universitat de Bar- celona
10. Dimiza MD, Triantaphyllou MV, Dermitzakis MD (2008) Seasonality and ecology of living coccolithophores in Eastern Mediterranean coastal environments (Andros Island, Middle Aegean Sea). Micropaleontology 54:18
11. Fujiki T, Taguchi S (2002) Variability in chlorophyll a specific absorption coefficient in marine phytoplankton as a function of cell size and irradiance. J Plankton Res 24:859–874. https://doi.org/10.1093/plankt/24.9.859
12. Gong X, Lembke-Jene L, Lohmann G, Knorr G, Tiedemann R, Zou JJ, Shi XF (2019) Enhanced North Pacific deep-ocean stratification by stronger intermediate water formation during Heinrich Stadial 1. Nat Commun 10:656. https://doi.org/10.1038/s41467-019-08606-2

13. Gordon HR, Brown OB, Evans RH, Brown JW, Smith RC, Baker KS, Clark DK (1988) A semianalytic radiance model of ocean color. J Geophys Res 93:10909. https://doi.org/10.1029/JD093iD09p10909
14. Harid R (2022) Coccolithophore blooms days detected by MODIS Level-2 data in Algiers bay between 2003 and 2018. Zenodo. https://doi.org/10.5281/zenodo.6612048
15. Harid R, Demarcq H, Keraghel M-A, Ait-Kaci M, Zerrouki M, Bachari N-E-I, Houma F (2022) Spatio-temporal variability of a chlorophyll-a based biomass index and influence of coastal sources of enrichment in the Algerian Basin. Cont Shelf Res 232:104629. https://doi.org/10.1016/j.csr.2021.104629
16. Hoepffner N, Sathyendranath S (1992) Bio-optical characteristics of coastal waters: absorption spectra of phytoplankton and pigment distribution in the western North Atlantic. Limnol Oceanogr 37:1660–1679. https://doi.org/10.4319/lo.1992.37.8.1660
17. Holligan PM, Charalampopoulou A, Hutson R (2010) Seasonal distributions of the coccolithophore, Emiliania huxleyi, and of particulate inorganic carbon in surface waters of the Scotia Sea. J Mar Syst 82:195–205. https://doi.org/10.1016/j.jmarsys.2010.05.007
18. Holligan PM, Viollier M, Harbour DS, Camus P, Champagne-Philippe M (1983) Satellite and ship studies of coccolithophore production along a continental shelf edge. Nature 304:339–342. https://doi.org/10.1038/304339a0
19. Iida T, Saitoh SI, Miyamura T, Toratani M, Fukushima H, Shiga N (2002) Temporal and spatial variability of coccolithophore blooms in the eastern Bering Sea, 1998–2001. Prog Oceanogr 55:165–175. https://doi.org/10.1016/S0079-6611(02)00076-9
20. Illoul H, Masó M, Alos J, Cros Miguel L, Morales-Blake A, Séridji R (2008) Potentially harmful microalgae in coastal waters of the Algiers area (Southern Mediterranean Sea). Cryptogam Algol 29:261–278
21. Jerome JH, Bukata RP, Miller JR (1996) Remote sensing reflectance and its relationship to optical properties of natural waters. Int J Remote Sens 17:3135–3155. https://doi.org/10.1080/01431169608949135
22. Lohrenz SE, Weidemann AD, Tuel M (2003) Phytoplankton spectral absorption as influenced by community size structure and pigment composition. J Plankton Res 25:35–61. https://doi.org/10.1093/plankt/25.1.35
23. Martin S, Seelye M (2004) An introduction to ocean remote sensing. Cambridge University Press
24. Merico A, Tyrrell T, Brown CW, Groom SB, Miller PI (2003) Analysis of satellite imagery for Emiliania huxleyi blooms in the Bering Sea before 1997. Geophys Res Lett 30. https://doi.org/10.1029/2002GL016648
25. Milliman JD (1993) Production and accumulation of calcium carbonate in the ocean: budget of a nonsteady state. Glob Biogeochem Cycles 7:927–957. https://doi.org/10.1029/93GB02524
26. Millot C, Taupier-Letage I, Benzohra M (1990) The Algerian eddies. Earth Sci Rev 27:203–219. https://doi.org/10.1016/0012-8252(90)90003-E
27. Mitchell C, Hu C, Bowler B, Drapeau D, Balch WM (2017) Estimating particulate inorganic carbon concentrations of the Global Ocean from ocean color measurements using a reflectance difference approach. J Geophys Res Oceans 122:8707–8720. https://doi.org/10.1002/2017JC013146
28. Moore CM, Suggett D, Holligan PM, Sharples J, Abraham ER, Lucas MI, Rippeth TP, Fisher NR, Simpson JH, Hydes DJ (2003) Physical controls on phytoplankton physiology and production at a shelf sea front: a fast repetition-rate fluorometer based field study. Mar Ecol Prog Ser 259:29–45. https://doi.org/10.3354/meps259029
29. Moore TS, Campbell JW, Dowell MD (2009) A class-based approach to characterizing and mapping the uncertainty of the MODIS ocean chlorophyll product. Remote Sens Environ 113:2424–2430. https://doi.org/10.1016/j.rse.2009.07.016
30. Moore TS, Dowell MD, Franz BA (2012) Detection of coccolithophore blooms in ocean color satellite imagery: a generalized approach for use with multiple sensors. Remote Sens Environ 117:249–263. https://doi.org/10.1016/j.rse.2011.10.001

31. Morel A, Prieur L (1977) Analysis of variations in ocean color1: ocean color analysis. Limnol Oceanogr 22:709–722. https://doi.org/10.4319/lo.1977.22.4.0709
32. NASA's Ocean Color Web (2019) Available online: http://oceancolor.gsfc.nasa.gov/. Accessed on 29 October 2019. [WWW Document]
33. Norme (2006) NF EN 15204, Normes Françaises et Européennes. ed, Normes nationales et documents normatifs nationaux. France
34. Nunes S, Perez GL, Latasa M, Zamanillo M, Delgado M, Ortega-Retuerta E, Marrasé C, Simó R, Estrada M (2019) Size fractionation, chemotaxonomic groups and bio-optical properties of phytoplankton along a transect from the Mediterranean Sea to the SW Atlantic Ocean. Sci Mar 83:87–109. https://doi.org/10.3989/scimar.04866.10A
35. Perrot L, Gohin F, Ruiz-Pino D, Lampert L (2016) Seasonal and interannual variability of coccolithophore blooms in the North East-Atlantic Ocean from a 18-year time-series of satellite water-leaving radiance (preprint). Remote Sen/Biol Proces/Surf/Shelf Seas. https://doi.org/10.5194/os-2016-13
36. Perrot L, Gohin F, Ruiz-Pino D, Lampert L, Huret M, Dessier A, Malestroit P, Dupuy C, Bourriau P (2018) Coccolith-derived turbidity and hydrological conditions in May in the Bay of Biscay. Prog Oceanogr 166:41–53. https://doi.org/10.1016/j.pocean.2017.12.008
37. Poulton AJ, Holligan PM, Charalampopoulou A, Adey TR (2017) Coccolithophore ecology in the tropical and subtropical Atlantic Ocean: new perspectives from the Atlantic meridional transect (AMT) programme. Prog Oceanogr 1995–2016(158):150–170. https://doi.org/10.1016/j.pocean.2017.01.003
38. Rabehi W, Guerfi M, Mahi H (2019) La baie d'Alger, un espace côtier prisé, entre pressions d'urbanisation et gouvernance territoriale. Geo-Eco-Marina 25:113–130. https://doi.org/10.5281/ZENODO.3609744
39. Raimbault P, Coste B, Boulhadid M, Boudjellal B (1993) Origin of high phytoplankton concentration in deep chlorophyll maximum (DCM) in a frontal region of the Southwestern Mediterranean Sea (algerian current). Deep-Sea Res I Oceanogr Res Pap 40:791–804. https://doi.org/10.1016/0967-0637(93)90072-B
40. Shutler JD, Land PE, Brown CW, Findlay HS, Donlon CJ, Medland M, Snooke R, Blackford JC (2013) Coccolithophore surface distributions in the North Atlantic and their modulation of the air-sea flux of CO_2 from 10 years of satellite Earth observation data. Biogeosciences 10:2699–2709. https://doi.org/10.5194/bg-10-2699-2013
41. Taupier-Letage I, Millot C (1988) Surface circulation in the Algerian basin during 1984. Oceanol Acta 9:79–85. https://archimer.ifremer.fr/doc/00267/37811/
42. Thierstein H, Young J (2004) Coccolithophores: from molecular processes to global impact. Springer. https://doi.org/10.1007/978-3-662-06278-4
43. Triantaphyllou MV, Ziveri P, Tselepides A (2004) Coccolithophore export production and response to seasonal surface water variability in the oligotrophic Cretan Sea (NE Mediterranean). Micropaleontology 50:127–144
44. Winter A (1994) Atlas of living coccolithophores. Coccolithophores:107–159

Multiscale Spatiotemporal NDVI Mapping of Salt Marshes Using Sentinel-2, Dove, and UAV Imagery in the Bay of Mont-Saint-Michel, France

Antoine Collin, Dorothée James, Antoine Mury, Mathilde Letard, Thomas Houet, Hélène Gloria, and Eric Feunteun

Abstract Salt marshes offer a plethora of ecosystem services such as biodiversity support, ocean–climate change regulation, ornithology recreo tourism or plant gathering by hand. They undergo significant worldwide losses due to their conversion into crop fields and to their spatial compression between rising sea levels and armored shorelines. Their management requires multiscale spatiotemporal analysis to detect interrelated patterns and processes. This research innovatively studies continuous salt marsh mapping, based on normalized difference vegetation index (NDVI) ranges, across three spatial and two temporal scales. Sentinel-2 (10 m), Dove (3 m), and unmanned airborne vehicle (UAV) (0.03 m) imagery were used to progressively refine spatial resolutions over dynamic areas (extending from hundreds, tens, and a couple of km^2, respectively). NDVI ranges from Sentinel-2 and Dove imagery were monitored with a lag of 5 and 4 years, respectively. Contrary to spaceborne imagery, UAV imagery lacked a near-infrared (NIR) band. The UAV NIR band was thus modelled (R^2_{NIR} = 0.98) using a three-layered neutral network (NN) prediction based on red, green, and blue reflectance imagery, itself calibrated/validated/tested by Dove imagery bands ($R^2_{\mathrm{red}} = 0.88$, $R^2_{\mathrm{green}} = 0.84$, and $R^2_{\mathrm{blue}} = 0.90$). The 100-fold increase in pixel size allowed to detect the decimeter-scale objects of salt marshes and tidal flats. The multiscale NDVI ranges were associated with

---

A. Collin (✉) · D. James · A. Mury · M. Letard · H. Gloria
Ecole Pratique des Hautes Etudes – PSL University, Coastal GeoEcological Lab, Dinard, France
e-mail: antoine.collin@ephe.psl.eu; dorothee.james@ephe.psl.eu; antoine.mury@etu.ephe.psl.eu; mathilde.letard@ephe.psl.eu; helene.gloria@ephe.psl.eu

T. Houet
Université of Rennes 2, CNRS LETG, Rennes, France
e-mail: thomas.houet@univ-rennes2.fr

E. Feunteun
Museum National d'Histoire Naturelle, Ecole Pratique des Hautes Etudes, CNRS BOREA, Dinard, France
e-mail: eric.feunteun@mnhn.fr

S. Niculescu (ed.), *European Spatial Data for Coastal and Marine Remote Sensing*, https://doi.org/10.1007/978-3-031-16213-8_2

microphytobenthos and topographically low, medium, and high salt marsh vegetation, including the opportunistic *Elymus* genus. The combination of the NDVI values derived from the Sentinel-2, Dove, and UAV imagery enabled to survey a region while detecting subtle features of salt marshes, providing an updated toolbox for managers.

Keywords High to very high temporal resolution · High to ultra-high spatial resolution · Spatiotemporal analysis · NDVI · Neural network modelling · NIR

1 Introduction

Salt marshes consist of tidal wetlands at the land–sea temperate interface in which complex interactions between shoots, roots, soil, and freshwater and seawater processes occur [19]. These critical habitats provide numerous ecosystem services, such as the biodiversity niche support (e.g., sentinel and umbrella plant species, benthic molluscs and crustaceans, insects, migratory and sedentary birds, and fish nursery within tidal channels), ocean–climate change regulation (hydrodynamic barrier against erosion and submersion, land fertilizer recycler, carbon sink [thus their nickname of "blue carbon," shared with mangroves and seagrasses]), ornithology recreo tourism, or plant gathering by hand [23, 31]. Since the Holocene with the birth and emergence of agriculture, these fertile areas were globally converted into crop fields. Nowadays, coastal eutrophication continues to the reduction of salt marsh extent and the deterioration in their ecological condition [11]. Despite these pressures, some of these ecosystems have remained stable in the face of sea level rise [20]. Elsewhere, significant losses of salt marshes have been observed, for example, in China [15], affecting the sediment budget and even tide [13]. In the context of global warming and associated increase in land and sea surface temperatures, poleward shifts of mangroves may entail possible competition with salt marshes in America and Australasia [34].

Given their considerable ecosystem services and pressures that may lead to deterioration and losses, salt marshes require frequent investigation of their biophysical envelope area and fragmentation rate. Regional monitoring using a finer-resolution imagery of their spatial patterns is necessary to quantify and determine the rate of change of their endogenous and exogenous drivers. Cartographically, salt marshes were represented by a dedicated symbol on maps since the beginning of the eighteenth century [14]. Following the ground triangulation of salt marshes over more than two centuries, the 1950s airborne photographs improved the geolocation of their areal extent along with the advent of landscape ecology [33]. The addition of airborne near-infrared (NIR) information led to the interpretation of salt marshes' net primary productivity [35], paving the way to the implementation of the NIR sensor on a spaceborne platform dedicated to vegetation mapping (Landsat-4, [16]). One of the vegetation indices, which uses a ratio of the difference between the red (R) and NIR bands with their sum, is the normalized difference vegetation index

(NDVI). It has been applied to Landsat-1 over the Great Plains of the United States of America since the 1970s [32]. The NDVI-based Landsat series have enabled the monitoring of salt marshes across four decades, provided with (resampled) 60 m to (native) 30 m pixel size [24]. More recently, the optical Sentinel-2 sensor, launched in 2015, enabled the NDVI to be computed at 10 m spatial resolution provided with a 5-day revisit at the equator (compared to an approximate 16-day revisit of Landsat-8 at the equator), offering valuable time series datasets to use in change detection of the extent and ecological condition of the salt marsh vegetation [36]. Since 2016, the Dove nanosatellites [29], each carrying a four-band optical sensor (RGB-NIR), embodied a technological leap given its daily revisit at 3 m pixel size [7].

The spaceborne observation of the extent and ecological condition of salt marshes has also benefited from an increase in both spatial and spectral resolution with the WorldView-3 sensor, providing five visible, three NIR and eight mid-infrared, bands at 0.3 m pixel size [4, 5]. Usually restricted to active very-fine-scale spatial resolution, such as airborne topographic and/or bathymetric Light Detection and Ranging (LiDAR) data [3], the height and species composition of the salt marsh vegetation could be mapped using neural network (NN) predictive modelling [4]. By providing a higher spatial resolution, the three-band sensor carried by the unmanned aerial vehicle (UAV) can capture natural-colored (RGB) imagery at an ultra-high resolution, ranging from 0.01 m to 0.1 m pixel size [25]. The disadvantage of using such consumer-grade UAVs may reside in the necessary addition of the NIR band to the main RGB camera, decreasing its relative cost-effectiveness. The predicted NIR band improved the separability of various types of vegetation, such as salt marshes, soils, sediments, and water types. A recent study has showed how to predict the NIR reflectance values from the existing RGB reflectance values using a fully connected NN spatially explicit model [8].

Because remote sensing is a nexus of trade-offs, today's spaceborne regional and high-spatial resolution imagery suffers from the coarseness of the grain size, while airborne local and ultra to very high spatial resolution imagery loses from the smallness of the scene. Even if those spatial specificities have been upgraded along with the temporal and spectral resolution, there is still no optimal sensor to spectrally discriminate submeter salt marshes' structural and functional patterns at the weekly interval.

This research aims to test an original multiscale, spatiotemporal continuous mapping approach, based on progressive NDVI spatial enhancement over the most dynamic areas, to:

- Monitor the change in extent and ecological condition of salt marshes at high spatial resolutions
- Predict the reflectance values of the NIR band for the UAV sensor, to enable the calculation of NDVI, at ultra-high spatial resolution

over one of the largest salt marsh zones that exist in Europe, namely, the Bay of Mont-Saint-Michel, France (Fig. 1).

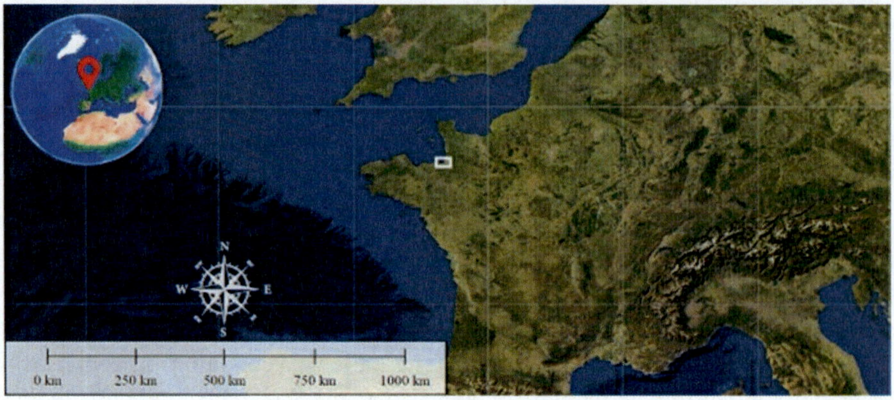

Fig. 1 Location of the study site hosting one of the largest extent of salt marsh in Europe: the Bay of Mont-Saint-Michel between Normandy and Brittany regions (France)

The objectives of the study are:

- To compare the changes in NDVI for salt marsh over a period of 4 and 5 years, using the 3 m Dove and 10 m Sentinel-2 imagery, respectively
- To produce 0.03 m spatially explicit NIR reflectance values and then derive NDVI values from the multi-layered NN learning developed with the UAV-borne senseFly RGB predictors and Dove responses

Spatial–temporal–spectral resolutions of the sensors and transferability of the modelling approach will be discussed in a context of salt marshes' monitoring and management.

2 Methodology

The methodological development of this research referred to a large-scale salt marsh area, very susceptible to encompass the spectral variability of various plants and soils, for the sake of transferability to other salt marshes. Technologically, Sentinel-2, Dove, and UAV sensors used in this study consist of the remote sensing sensors that are increasingly used by scientists and managers.

2.1 Study Area

Salt marshes of the Bay of Mont-Saint-Michel lie along Normandy and Brittany in France (48°38′05″N; 1°30′36″W). Covering an extent of dozens of km^2 in the highest part of the intertidal area (only submerged during high spring tides), these tidal coastal ecosystems receive freshwater influence from the three main rivers in the east (the Sée, Sélune, and Couesnon rivers) and are subject to a clockwise gyre in the west (Fig. 2). That spatial heterogeneity, therefore, ranges from an estuarine ecosystem in the east, which is hydrologically very dynamic, to a bay system in the west, which, in comparison, is much calmer. Those salt marshes are classified as a natural heritage site by the UN Educational, Scientific, and Cultural Organization (UNESCO) in 1979, and benefit from European protection as Natura 2000 sites for both "Habitats" and "Birds" directives. Birdlife International also describes the bay and its salt marshes as an "important zone for bird conservation." On a national scale, they are inventoried and listed as "natural zones for ecology, flora, and fauna."

Those salt marshes host extensive human activities, ranging from livestock farming to hunting of waterbirds, for example, of which the impacts and pressures are evident from the reticulated pathways and artificial shallow water bodies across the area (Fig. 3).

2.2 *Imagery Source and Processing*

This research adopts a top-down approach by progressively focusing on the most dynamic salt marshes with the simultaneous increase in pixel size and decrease in scene size. This could be deemed as a downscaling from 10 m Sentinel-2 to 3 m Dove, in turn, to 0.03 m UAV imagery. The continuous mapping of this multiscale study is facilitated by ranges of NDVI values derived for these images.

Fig. 2 Sentinel-2 RGB values of the Bay of Mont-Saint-Michel, France, acquired on (**a**) 18 March 2016 and (**b**) 1 April 2021, respectively

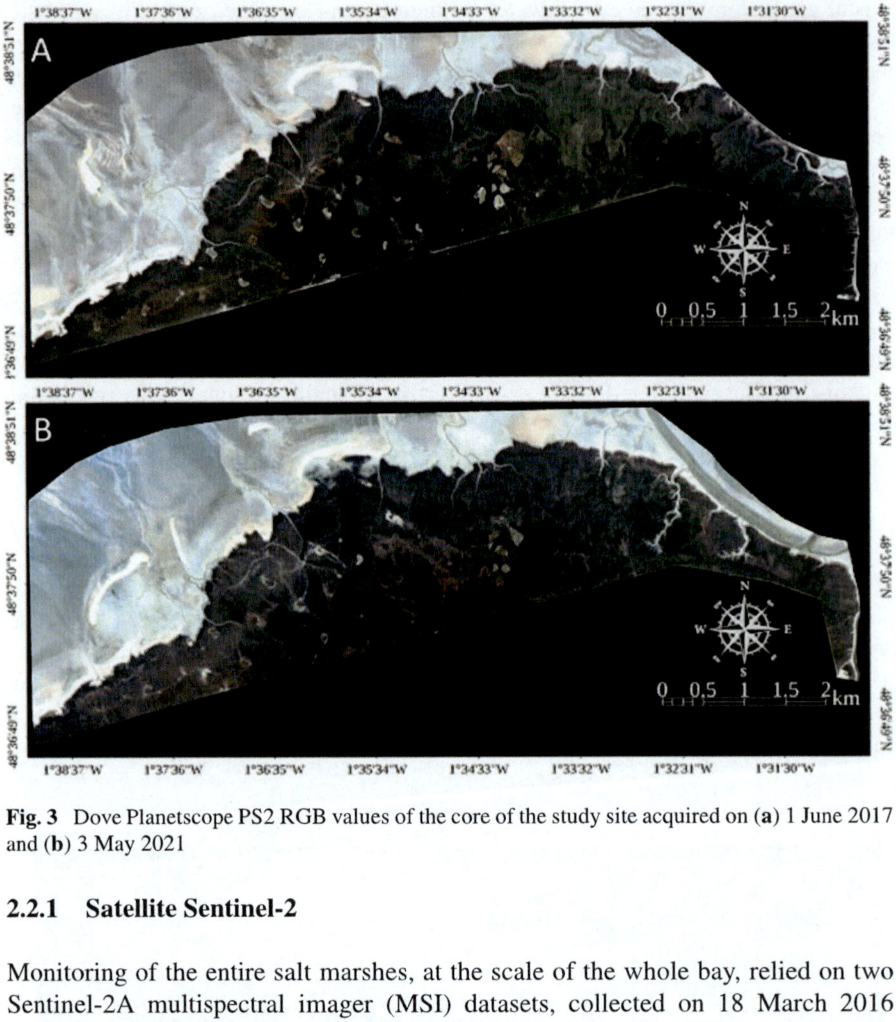

Fig. 3 Dove Planetscope PS2 RGB values of the core of the study site acquired on (**a**) 1 June 2017 and (**b**) 3 May 2021

2.2.1 Satellite Sentinel-2

Monitoring of the entire salt marshes, at the scale of the whole bay, relied on two Sentinel-2A multispectral imager (MSI) datasets, collected on 18 March 2016 (Fig. 2a) and 1 April 2021 (Fig. 2b). The *B* (492 ± 33 nm), *G* (560 ± 18 nm), *R* (665 ± 15 nm), and NIR (833 ± 53 nm) wavebands were processed at 10 m pixel size. The NDVI values were calculated from the R and NIR bands.

Imagery was downloaded as a L1C product, which is geometrically and radiometrically corrected to the top of atmosphere (TOA) reflectance. Using the Sentinel-2 toolbox (SNAP v.8), the atmosphere effect was calculated to convert the TOA reflectance in bottom of atmosphere (BOA) reflectance (i.e., the L2A product).

NDVI values were then computed for both dates following Eq. 1 [32]:

$$\mathrm{NDVI} = \left(\mathrm{NIR}_{\mathrm{BOAref}} - \mathrm{Red}_{\mathrm{BOAref}}\right) / \left(\mathrm{NIR}_{\mathrm{BOAref}} + \mathrm{Red}_{\mathrm{BOAref}}\right) \quad (1)$$

where $\mathrm{NIR}_{\mathrm{BOAref}}$ and $\mathrm{Red}_{\mathrm{BOAref}}$ corresponded to the Sentinel-2A bands centered at 833 and 665 nm, respectively.

The 5-year diachronic analysis established the per-pixel difference of the NDVI values from each image. These image dates were chosen to be the most distant while being captured at the closest seasonal time, here in the early spring.

2.2.2 Nanosatellite Dove

Based on large-scale Sentinel-2A information, the largest and most dynamic salt marsh area (centered at 48°38'N; 1°35'W on Fig. 2) was examined at a finer scale using hypertemporal Dove imagery [7], from Planet Labs. Two Dove imagery, provided with the Planetscope PS2 camera, were acquired on 1 June 2017 (Fig. 3a) and 3 May 2021 (Fig. 3b). The B (485 ± 30 nm), G (545 ± 45 nm), R (630 ± 40 nm), and NIR (820 ± 40 nm) wavebands were examined at 3 m spatial resolution. Geometric and radiometric corrections were applied to obtain the orthorectified BOA reflectance. NDVI values was also calculated for each date, and then the 4-year difference was mapped.

A third Dove image (Fig. 4a) was retrieved for the most extensive and dynamic salt marsh area along the coast (centered at 48°37'50"N; 1°36'35" on Fig. 3) to serve as spectral responses of the NN predictive modelling, involving RGB predictors from UAV. The image was taken on 11 July 2020 to be the closest to the UAV flight conducted on 8 July 2020 (Fig. 4b), which featured the same spectral and spatial specificities and geometric and radiometric corrections, as previously mentioned.

2.2.3 Unmanned Aerial Vehicle

By investigating the medium-scale Dove Planetscope PS2 information, the changing zone was further analyzed at the local scale using the ultra-high-resolution UAV-borne senseFly "sensor optimized for drone application" (SODA) sensor. This 5472 × 3648-pixel camera was mounted on a fixed-wing eBee UAV [26], whose flight mission was planned at 120 m height above mean sea level corresponding to a 0.03 m pixel size on the ground. The survey took place on 8 July 2020, 3 days before the Dove acquisition.

As a conventional visible camera, the SODA sensor captured RGB wavebands in the form of digital numbers (DN) coded into eight bits.

2.3 Predicting the NIR at the UAV Scale

The NIR constitutes a compulsory term to solve for the NDVI formula. That waveband, slightly longer than the visible gamut, exists for the Sentinel-2 and Dove satellite sensors, but not for the UAV camera. By implementing a nonlinear regression, the Dove NIR BOA reflectance values' response could be predicted from the UAV RGB DN predictors, using artificial NN models [8].

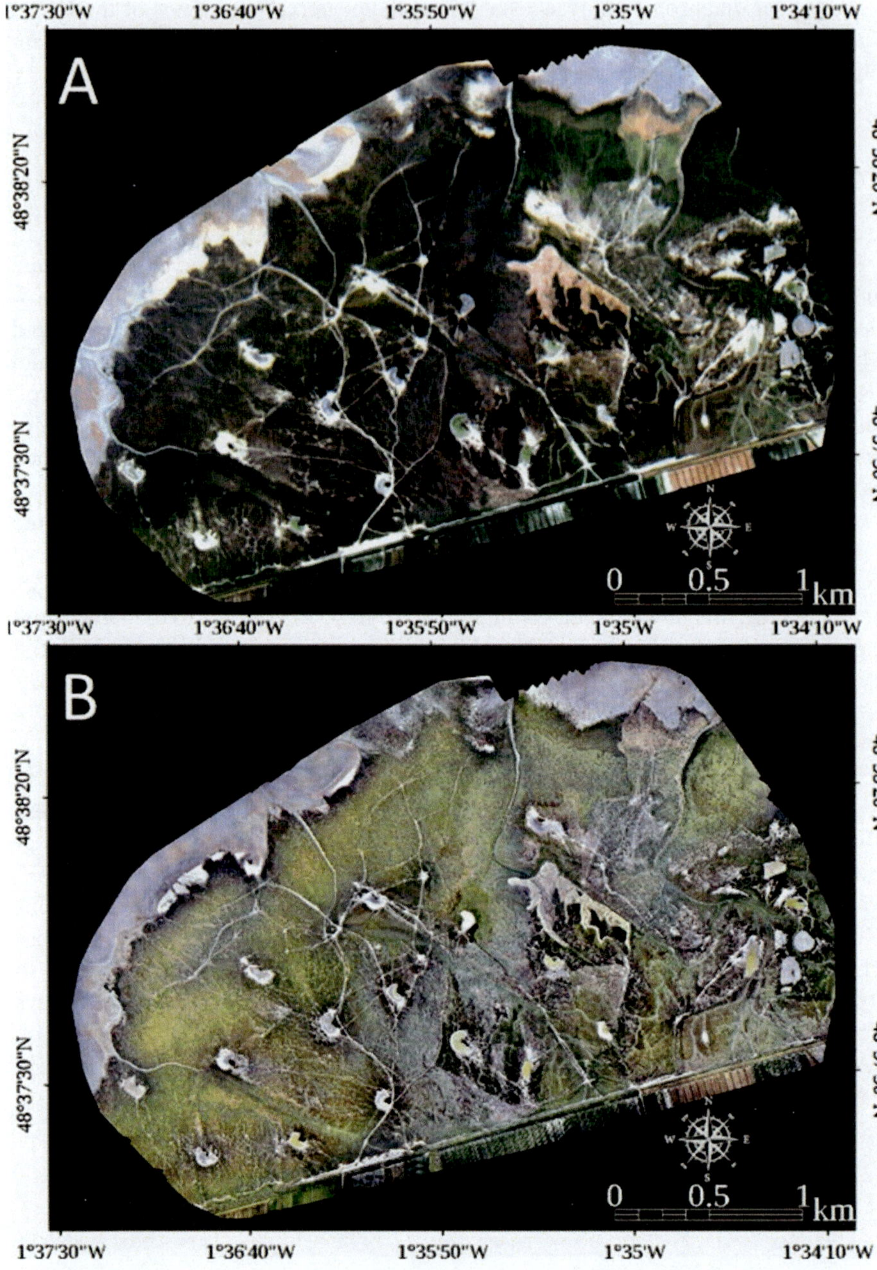

Fig. 4 RGB values derived from (**a**) Dove Planetscope PS2 imagery, acquired on 11 July 2020 and (**b**) unmanned airborne vehicle SODA, acquired on 8 July 2020

2.3.1 Two-Scaled Spectral Datasets

The potential to accurately simulate the NIR BOA reflectance for the UAV sensor depends on the sampling strategy (Fig. 5).

The proposed procedure started with a stack of the Dove NIR-RGB BOA reflectance with the UAV RGB DN using the co-registration. For each spectral satellite reflectance band to be predicted, a relevant pixel sampling was carried out:

- Slicing the spectral band into 75 equal reflectance ranges
- Saving the sliced areas as regions of interest
- Calculating and extracting the mean values for the four satellite sensors and three UAV bands per slice
- Randomly dividing the output matrix into 25 training, 25 test, and 25 validation samples

2.3.2 Neural Network Regression

The estimation of the reflectance values in the NIR at BOA from the UAV RGB DN required the predictions of the *R*, *G*, and *B* BOA reflectances. Based on the band-specific stratified sampling, the NN model built a fully connected one-layer perceptron, whose (hidden) layer was composed of a single (hidden) neuron (or node) [17]. For each neuron, a sigmoid activation function, defined as a hyperbolic tangent function (TanH), was generated as a linear regression of the RGB predictors (Eq. 2):

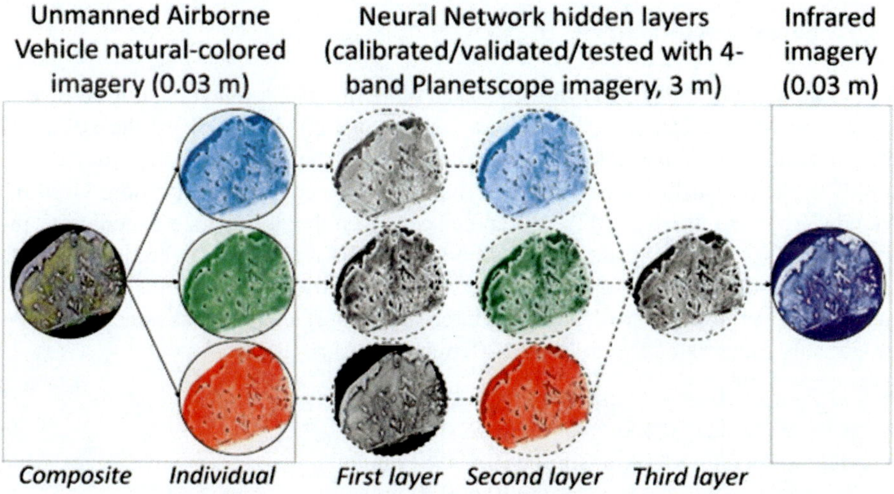

Fig. 5 Flowchart describing how a 3 m satellite Dove Planetscope PS2 near-infrared imagery (bottom of atmosphere reflectance) can be spatially enhanced at 0.03 m pixel size, using 0.03 m unmanned airborne vehicle natural-colored imagery (digital numbers) and a three-layer neural network calibrated/validated/tested by the satellite Planetscope blue–green–red–near-infrared imagery

$$\mathrm{NN}(X_i) = w_i X_i + C \tag{2}$$

where NN(X_i) is the predicted BOA reflectance of the X_i spectral band X_i, w_i is the sigmoid activation function of the DN spectral band, and C is an offset (see the first hidden layer in Fig. 5).

Once the RGB datasets were properly estimated at the BOA reflectance level (see the second layer in Fig. 5), the NIR BOA reflectance could have been modelled using a hidden layer with a single neuron linked with the three predicted RGB BOA reflectance values (see the third layer in Fig. 5).

3 Results

The innovative multiscale continuous mapping of salt marshes in the Bay of Mont-Saint-Michel relied, first, on the diachronic analysis of the 5-year and 4-year change in the NDVI using 10 m Sentinel-2 and 3 m Dove imagery, respectively, and second, on the prediction of the NDVI, using the 0.03 m UAV along with a NIR NN modelling.

3.1 *Changes in the Normalized Difference Vegetation Index Values*

3.1.1 Satellite Sentinel-2

The calculation of the NDVI products based on Sentinel-2A imagery showed divergences between 18 March 2016 (Fig. 6a) and 1 April 2021 (Fig. 6b) at the extent of the full study area. The sea level was different due to the variation in the tide schedule. Crop fields, bushes, trees, and salt marshes were conspicuously more vivid in March 2016 than in April 2021, as can be seen from the dominance of red color in the March 2016 image. In fact, the number of NDVI ranges >0 covered an area of 241.6 km^2 and 214.4 km^2 for March 2016 and April 2021, respectively. By examining the NDVI ranges >0 with a 0.2 lag interval, the surface area was generally geographically more extensive in March 2016 than in April 2021:

- 111.2 km^2 versus 92.0 km^2 for the [0–0.2] range
- 53.9 km^2 versus 50.4 km^2 for the [0.2–0.4] range
- 53.2 km^2 versus 54.1 km^2 for the [0.4–0.6] range
- 23.3 km^2 versus 17.8 km^2 for the [0.6–0.8] range

Focused on salt marshes (see the masking in Fig. 7), the distribution of the NDVI values varied between March 2016 and April 2021. NDVI values <0 spanned over smaller areas in March 2016 (1.1 km^2) than in April 2021 (1.6 km^2). On the contrary,

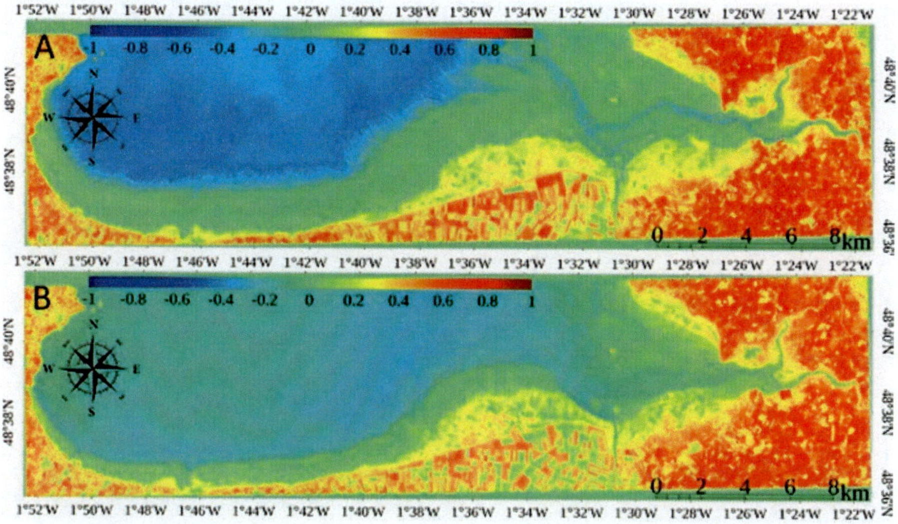

Fig. 6 NDVI values of the study site derived from the Sentinel-2 imagery acquired on (**a**) 18 March 2016 and (**b**) 1 April 2021, respectively

the NDVI values >0 extended over a slightly larger overall area in March 2016 (59 km^2) than in April 2021 (58.5 km^2). However, this average concealed trends at the NDVI range level with a predominance in 2021 compared to 2016, except for the medium positive NDVI range:

- 27.4 km^2 versus 25.4 km^2 for the [0–0.2] range
- 24.6 km^2 versus 29.7 km^2 for the [0.2–0.4] range
- 6.2 km^2 versus 3.7 km^2 for the [0.4–0.6] range
- 0.3 km^2 versus 0.2 km^2 for the [0.6–0.8] range.

The difference in NDVI between the early springs of 2021 and 2016 (Fig. 7) was characterized by a slight increase in negative and very high positive [0.6–0.8] NDVI values, a stronger augmentation for low [0–0.2] and high positive [0.4–0.6] NDVI values, and a marked decline in medium positive [0.2–0.4] NDVI values.

The white rectangle on Fig. 7 outlined the area provided with the deepest contrast in the salt marsh 2016–2021 change, which was further examined with the Dove imagery.

3.1.2 Nanosatellite Dove

The calculation of the NDVI products based on Dove Planetscope PS2 revealed distinctions between 1 June 2017 (Fig. 8a) and 3 May 2021 (Fig. 8b) on the medium scale of the most dynamic salt marsh. The NDVI values seemed, at first sight, to decrease in that 4-year period, as attested by the loss in the orange color in May

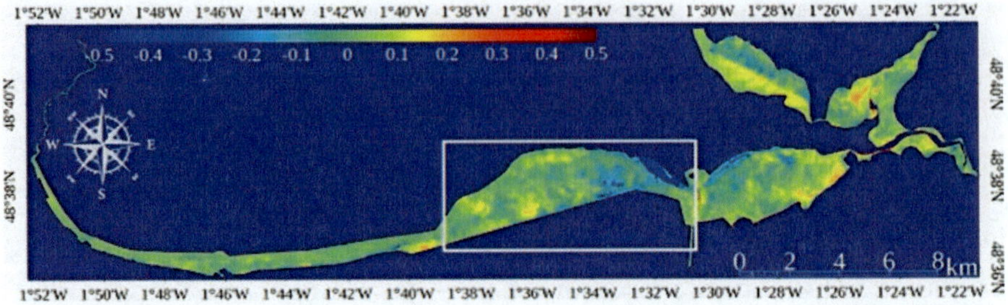

Fig. 7 Difference in the NDVI values of the Sentinel-2 imagery between 1 April 2021 and 18 March 2016. The white rectangle identifies the dynamic area that was further investigated with the Dove Planetscope PS2 imagery

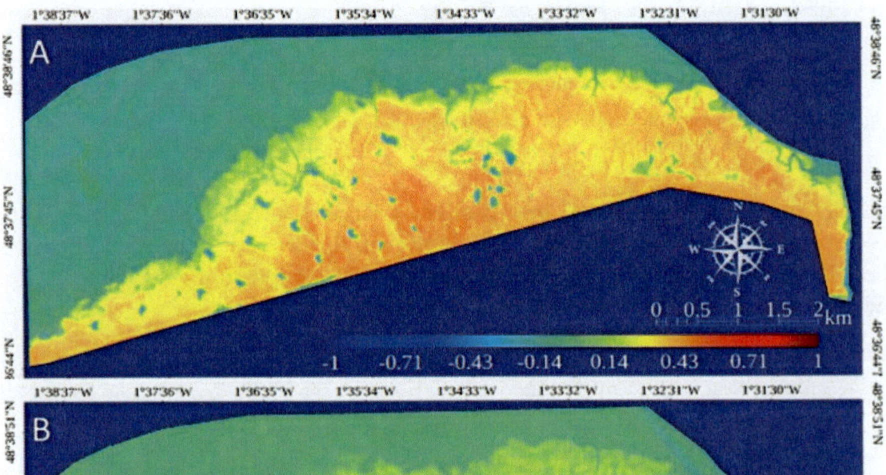

Fig. 8 Dove Planetscope PS2 NDVI values of the core of the study site acquired on (**a**) 1 June 2017 and (**b**) 3 May 2021, respectively

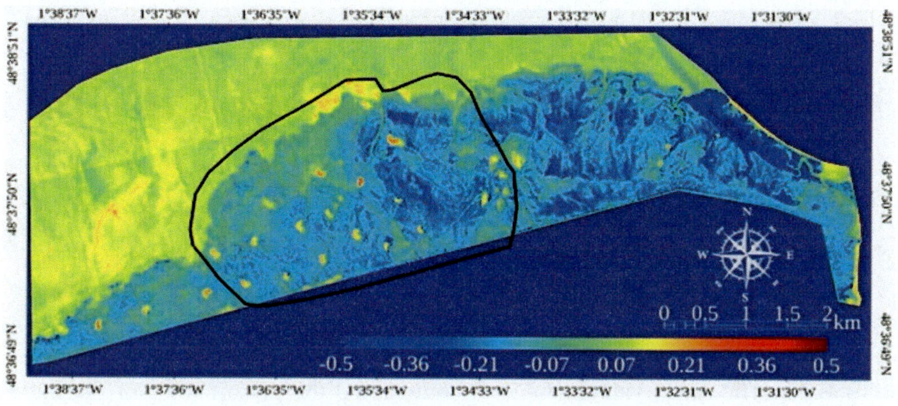

Fig. 9 Difference in the NDVI values derived from the Dove Planetscope PS2 imagery between 3 May 2021 and 1 June 2017. The black polygon identifies the salt marsh area that was surveyed by an unmanned airborne vehicle on 8 July 2020

2021. The easternmost salt marsh zone has eroded to the benefit of a river channel (48°38'N; 1°31'30"W). In contrast, the central western front has evidently prograded (48°38'N; 1°36"W).

NDVI ranges <0 decreased (from 11.2 km2 in June 2017 to 11.1 km^2 in May 2021), whereas NDVI ranges >0 waxed (from 14.55 km^2 in June 2017 to 14.64 km^2 in May 2021). By examining NDVI ranges >0 at the 0.2 lag, the surface area was much wider in 2021 than in 2017 for the [0–0.2] range (10.2 km^2 versus 1.7 km^2) but was more cramped for both the [0.2–0.4] range (4.4 km^2 versus 10.4 km^2) and the [0.4–0.6] range (0.005 km^2 versus 2.46 km^2).

The difference in NDVI between the late springs of 2021 and 2017 (Fig. 9) was characterized by a slight rise in negative [0.6–0.8] NDVI values, the greatest increase for low positive [0–0.2] NDVI values, and a tangible decrease in the medium [0.2–0.4] and high [0.4–0.6] positive NDVI values.

The black rectangle in Fig. 9 delineated the hybrid area of interest provided with obvious losses and gains in NDVI values in the salt marsh during the period from 2017 to 2021. That zone was further investigated with the ranges of NDVI values derived from the UAV imagery.

3.2 *Prediction of the Normalized Difference Vegetation Index Values*

3.2.1 Unmanned Aerial Vehicle Near-Infrared

The study of NDVI spatial patterns was not trivial and required one to be able to predict a NIR band at a fine scale for the UAV sensor. Based on a previous development established on NN predictive modelling of the NIR reflectance value from

Table 1 Results of the neural network predictive modelling for each spectral band as a function of the randomly divided datasets

Spectral band	Test	
	R^2	RMSE
Blue	0.90	51.26
Green	0.84	68.55
Red	0.88	82.70
Near-infrared	0.98	162.73

UAV RGB predictors [8], the UAV NIR BOA reflectance values were predicted from the UAV RGB BOA reflectance values, themselves predicted from the Dove RGB BOA reflectance values. Table 1 summarized the performance of the three individual RGB predictions, as well as the final NIR prediction. The RGB BOA reflectance values were considered well predicted (by the one-neuroned one-layered model) because they achieved coefficients of determinations (R^2) ranging from 0.84 to 0.90. The NIR BOA reflectance values were very well predicted by its three-neuroned one-layered NN model, with an R^2 attaining a maximum value of 0.98.

The predicted values of the RGB NN models were combined to predict the NIR BOA reflectance values (using the Dove Planetscope PS2 imagery, Fig. 10a), following the formula:

$$\begin{aligned} \mathrm{NIR}_{\mathrm{BOAref}} &= 2755.53 + 2025.37 \times \\ &(\tan H\left(0.5 \times \left(12.12 - 0.02 \times B_{\mathrm{BOAref}} + 0.01 \times G_{\mathrm{BOAref}} - 0.02 \times R_{\mathrm{BOAref}}\right)\right) \end{aligned} \quad (3)$$

where all NIR, R, G, and B variables correspond to the predicted BOA reflectance spectral variables.

The equation was then applied to the UAV pixel to predict the NIR BOA reflectance at a higher spatial resolution (Fig. 10b). Although the general spatial patterns were qualitatively similar between the 3 m and 0.03 m NIR BOA products, such as for artificial water bodies for hunting and tidal flats, the 100-fold magnification details calculated for the salt marsh allowed elucidating the complex network of sheep and hunters' pathways and also the higher vegetation reflectance values at the front and around the water bodies. Conversely, the more subtle reflectance signatures allowed identifying Dove homogeneous vegetated areas that were decomposed in a larger diversity of plant reflectance values, thus taxonomy and/or phenology.

The creation of the NIR BOA reflectance band on the Dove (Fig. 10a) and UAV (Fig. 10b) made the NDVI calculation possible (Fig. 11a, b, respectively). At the level of the salt marsh area, the spatial variability of the NDVI associated with the innovative UAV-based imagery was obviously much stronger than this tied to the Dove-based data, labeled with larger even areas. To highlight the great information enhancement derived from the UAV, a zoom-in was realized (the small black rectangles in Fig. 11), incontrovertibly contrasting the homogeneity of the 3 m Dove NDVI values, hardly fluctuating, with the tremendous complexity of the 0.03 m UAV NDVI values.

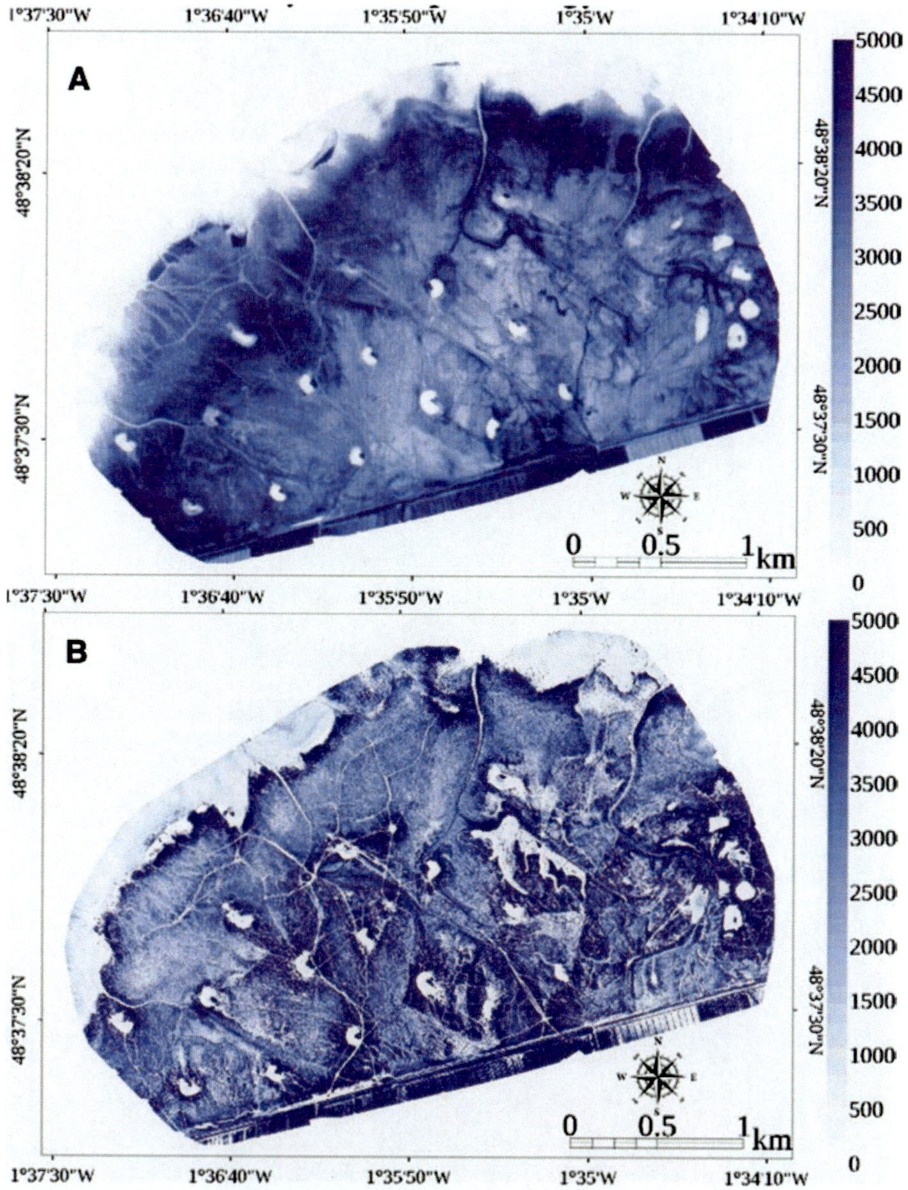

Fig. 10 NIR values derived from (**a**) Dove Planetscope PS2 bottom of atmosphere reflectance (BOA), acquired on 11 July 2020, and (**b**) predicted BOA reflectance for the UAV, from the 8 July 2020 acquisition

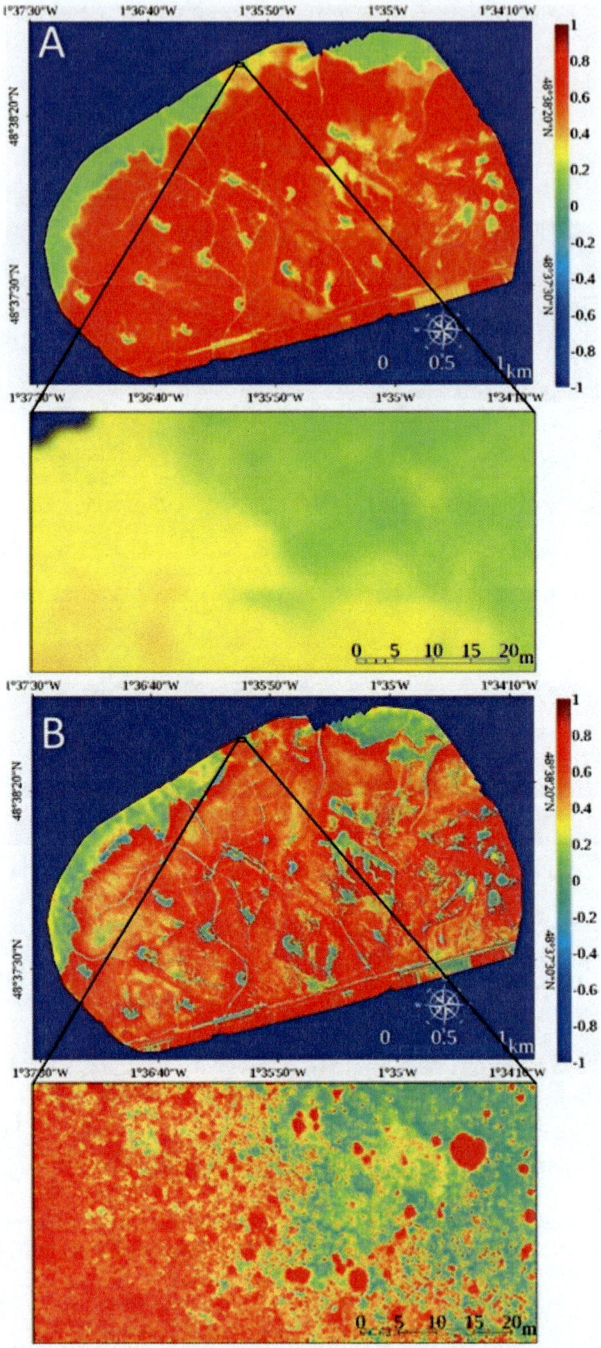

Fig. 11 NDVI values derived from the (a) Dove 3 m Planetscope PS2 imagery, acquired on 11 July 2020, and (**b**) SODA 0.03 m UAV, collected on 8 July 2020. The black rectangles correspond to close-ups

Quantitatively, the number of NDVI ranges and the surface area covered within ranges (defined at the 0.2 lag) conspicuously varied from the Dove to UAV products. The Dove dataset ranged from [−0.2–0] to [0.8–1.0], encapsulating six ranges, while the UAV comprised seven ranges, adding the [−0.4−0.2] to the previous ones. Even if the relative distribution within common ranges followed the same tendency for both datasets, the UAV detected more negative, medium, and extreme positive NDVI ranges at the detriment of the high and very high positive ranges.

4 Discussion

4.1 Salt Marsh Spatiotemporal Patterns

4.1.1 Satellite-Based Salt Marsh Spatial Analysis

According to the results derived from the 10 m Sentinel-2 NDVI classification, the surface covered by vegetation tends to average 59 km^2 for the whole Bay of Mont-Saint-Michel, thus including the coastal ecosystems of Brittany and Normandy. That area encompassed subareas associated with diverse NDVI ranges, such as low, medium, high, and very high values. Some of the low values of NDVI [0–0.2] might stem from the phytoplankton flourishing on the narrow tidal flats and channels [38], wandering in salt marsh meadows. A large but not exclusive part of the medium NDVI range [0.2–0.4] could emerge from the topographically low salt marsh plant genera, such as *Spartina*, *Salicornia*, and *Suaeda* [12, 30]. Likewise, the high NDVI range [0.4–0.6] partly arose from the topographically medium salt marsh flagship species *Halimione portulacoides* [5]. Finally, the very high NDVI range [0.6–0.8] might contain the spectral signatures of the topographically high salt marsh genera, such as *Puccinellia*, *Festuca*, *Aster*, and *Limione* [1, 27]. In that latter NDVI range could also be integrated the genus *Elymus* [37], which tends to thrive over the topographically medium and high salt marsh at the detriment of traditional species and genera [39]. That *Elymus* expansion, between 1984 and 2013, within the Bay of Mont-Saint-Michel, was hypothesized as an ecological response of the watershed change in agricultural land use, thus in eutrophication (especially nitrates).

The previous authors established that salt marshes extended over 42 km^2 in 2013 based on a thorough photo-interpretation. Our NDVI products overestimated those findings by 18 km^2. However, by keeping medium, high, and very high NDVI ranges (those actually related to salt marsh vegetation) and simply removing the half of the surface covered by the low NDVI range [0–0.2], which was very likely to merge microphytobenthos and sparse topographically low salt marsh vegetation, the results reached 46 km^2.

On the scale of the most dynamic salt marsh, the 3 m Dove NDVI mapping, issued from slicing in 2017, clearly shed light on the predominance of the negative and positive medium NDVI ranges, compared to the low and high NDVI ranges. Therefore, the enhancement of spatial resolution could better discriminate the microphytobenthos from the topographically low salt marsh vegetation, provided that the salt marsh analyzed by Dove was representative of all salt marshes examined by Sentinel-2.

Inaccuracies derived from the manual outline of older studies and our NDVI continuous indicator could be further overcome by implementing a supervised classification at a very high resolution using machine learning [4], including deep learning [9].

4.1.2 Satellite-Based Salt Marsh Temporal Change

At the bay scale, the change in NDVI values' ranges between 2016 and 2021 underlined a trend of salt marsh expansion around the three river mouths (east), a diminution south of Mont-Saint-Michel, and a development slightly westward (see Fig. 7). The latter erosion/accretion salt marsh movement could be explained by the flushing effect of the recently built (2009) dam just south of Mont-Saint-Michel. The overall salt marsh progradation, during those 5 years, forms the recent part of an expansion process, observed since the first aerial photographs in 1947: 2021 (46 km^2, our study), 2013 (42 km^2, [39]), 1996 (24.2 km^2, [2]), 1980 (21.8 km^2, [2]), and 1947 (12.9 km^2, [2]). The time series assessment at that bay scale will be duly enriched with the high temporal resolution provided by the 5-day 10 m Sentinel-2 to identify the stability or the variability in the velocity of the salt marsh general progradation and local erosion/accretion, facing anthropogenic changes [22].

The diachronic analysis of the 5-year (Sentinel-2) NDVI time series has unveiled that the early spring salt marshes at the bay scale gained topographically low ([0–0.2], +2.1 km^2) and high ([0.4–0.6], +2.5 km^2; [0.6–0.8], +0.1 km^2) salt marsh vegetation, but lost the medium ([0.2–0.4], −5.1 km^2) one. This pattern might be explained by the trend in *H. portulacoides* loss with the benefit of the low salt marsh, which prograded, and of the high, even very high vegetation, *Elymus*, putatively boosted by eutrophication.

The NDVI change of the representative salt marsh between late spring 2017 and middle spring 2021 (Dove results) was also characterized by an increase in low salt marsh vegetation ([0–0.2], +8.5 km^2), a decrease in medium salt marsh vegetation ([0.2–0.4], −6 km^2), but a contrasting decline in high vegetation ([0.4–0.6], −2.5 km^2), compared to Sentinel-2 results. This reduction in high vegetation could be corroborated with the fact that the focused salt marsh did not well represent the spread of *Elymus* insofar as the studied area was relatively remote from the three rivers likely to convey the excessive nitrate inputs. The variation in the high vegetation associated with high NDVI values could also be due to the phenological relative distance occurring between late spring (1 June 2017) and middle spring (3 May 2021).

The increase in low salt marsh vegetation could also be reinforced by the natural landward migration of the crescent-shaped sandy dunes that are visible in Fig. 9 (west). Originating from a 2 km^2 seaward biogenic reef (located in the lower intertidal area, close to the subtidal limit), erected by the honeycomb worm *Sabellaria alveolata* [6], sediment pulses of calcareous sand are actuated by the combination of stormy waves and strong tidal currents, crossing the intertidal area on average of 30 years [10]. When the biogenic sandy dunes get closer to the salt marsh forefront (hundreds of meters), the hydrodynamic conditions become calmer [27], which triggers an increase of rate sedimentation, in turn, facilitating the colonization of the pioneer salt marsh species, such as *Spartina*, *Salicornia*, and *Suaeda* genera. Like with Sentinel-2, the very high temporal resolution Dove satellites embody key imagery to build continuous time series that have the potential to neatly link the ecology of the most extensive animal reef construction in Europe, seaward, with one of the broadest vegetal meadows in Europe, landward, through a mobile landform.

4.2 Modelling the Extent of Salt Marsh at an Ultra-High Resolution

4.2.1 Spectral Predictive Modelling at Ultra-High Resolution

The great performance of the predicted NN modelling of the NIR BOA reflectance values helped to produce the NDVI at an unprecedented spatial resolution over salt marshes of the Bay of Mont-Saint-Michel. That scientific advancement enabled both the detection of an NDVI range [−0.4–0.2], invisible to Dove imagery, and the better pixel distribution in the NDVI ranges, by eroding the high and very high positive ranges and by meliorating the negative, medium, and extreme positive ranges: [0–0.2], −1.2 ha; [0.2–0.4], 29.5 ha; [0.4–0.6], −85.8 ha; [0.6–0.8], −82.6 ha; [0.8–1.0], +36.9 ha. Those refinements will straightforwardly enhance the diagnosis of the ecological structuration and functioning of salt marshes, laying the foundation of a sound spatialized ecological state indicator.

That indicator would be even more robust if it relied on additional spectral bands, such as the red edge (RE), "squeezed" between the *R* and the NIR, which is encompassed in the SuperDove PS2.SD satellite imagery. Furthermore, the SuperDove PSB.SD imagery even captures the purple, called coastal blue [431–452 nm], *B* [465–515 nm], two *G* [513–549 and 547–583 nm], yellow [600–620 nm], *R* [650–680 nm], RE [697–713 nm], and NIR [845–885 nm] bands. Along with the 0.3 m six-band Pléiades Neo and the 0.3 m eight-band WorldView-2 and WorldView-3, the 3 m PSB.SD imagery modelled at the UAV scale holds great promise to create optical multispectral ultra-high resolution imagery at low cost. An even cost-efficient alternative would be to downscale the visible and NIR spectral bands tied to Sentinel-2 at the UAV pixel size, while bearing in mind that the scaling process will not shift imagery from 3 m (or 0.3 m) to 0.03 m, but from 10 m to 0.03 m, which could lie on a strong assumption of the spectro-spatial continuity.

4.2.2 Predictive Modelling of Vegetation Height at Ultra-High Resolution

The NN predictive modelling of the NIR and other spectral optical bands will aid in elaborating ecological indicators such as vegetation, soil, or water continuous indices. Calibrated and validated by topographic LiDAR height, NN modelling even successfully predicted salt marsh vegetation height and elevation [4]. Despite its reference status to monitor coastal elevation [28] and species composition [3], an airborne LiDAR survey remains costly. Spaceborne LiDAR, such as the topo-bathymetric ICESat-2, could be a much more affordable solution to sample the height and elevation of salt marsh, while keeping in mind that the footprint approximately reaches 10 m diameter [21]. Otherwise, the height and elevation could also be calculated using the photogrammetric procedure with the UAV planning survey [18]. In addition to the multispectral UAV-scaled BOA reflectance bands (see above), the information of vegetation height and elevation will have a great potential to provide decisive cues to classify the salt marsh at the species level.

5 Conclusion

This paper sheds light on the NDVI-based salt marsh mapping using a spatiotemporal multiscale process based on state-of-the-art remote sensing technologies, from the 10 m Sentinel-2A MSI through the 3 m Dove Planetscope PS2 to the 0.03 m UAV imagery.

The Sentinel-2 and Dove multispectral BOA reflectance values helped to monitor NDVI ranges at the bay (hundreds of km^2) and the salt marsh (tens of km^2) scales, respectively. Their high and very high temporal resolution enabled them to build diachronic analysis across five (2016–2021) and four (2017–2021) years, respectively. Findings of both scales were discussed in the continuity of older studies, as well as hydrological, hydrodynamic, ecological, and socio-economic drivers. The NDVI ranges were discussed to discriminate microphytobenthos and topographically low, medium, and high salt marsh vegetation, including the opportunistic *Elymus* genus.

NDVI values were very satisfactorily modelled at the UAV scale (a few km^2) using a three-layer NN prediction, providing the final NIR and the intermediate RGB BOA reflectance values, calibrated/validated/tested by the Dove BOA reflectance values ($R^2_{NIR} = 0.98$, $R^2_{red} = 0.88$, $R^2_{green} = 0.84$, and $R^2_{blue} = 0.90$). The 100-fold gain in pixel size not only allowed to seize the decimeter-scale objects of salt marshes and tidal flats but also allowed to better distribute the NDVI values, by increasing the negative, medium, and extreme positive values, which were underestimated, and by decreasing the high and very high positive values, which were overestimated by the 3 m Dove BOA reflectance values.

Acknowledgments The authors acknowledge the European Space Agency, Planet, and French Office for the Biodiversity for their support in image acquisition. The anonymous reviewer and Ninon Powers helped tremendously in improving the content and style of the manuscript.

References

1. Belluco E, Camuffo M, Ferrari S, Modenese L, Silvestri S, Marani A, Marani M (2006) Mapping salt-marsh vegetation by multispectral and hyperspectral remote sensing. Remote Sens Environ 105(1):54–67
2. Bonnot-Courtois C, Fournier J, Dréau A (2004) Recent morphodynamics of shell banks in the western part of the Bay of Mont-Saint-Michel (France)/Morphodynamique actuelle des bancs coquilliers dans la partie occidentale de la baie du Mont-Saint-Michel (France). Géomorphol Relief Process Environ 10(1):65–79
3. Collin A, Long B, Archambault P (2010) Salt-marsh characterization, zonation assessment and mapping through a dual-wavelength LiDAR. Remote Sens Environ 114(3):520–530
4. Collin A, Lambert N, Etienne S (2018) Satellite-based salt marsh elevation, vegetation height, and species composition mapping using the superspectral WorldView-3 imagery. Int J Remote Sens 39(17):5619–5637
5. Collin A, Lambert N, James D, Etienne S (2018) Mapping wave attenuation induced by salt marsh vegetation using WorldView-3 satellite imagery. Rev Invest Marina 25(2):67–69
6. Collin A, Dubois S, Ramambason C, Etienne S (2018) Very high-resolution mapping of emerging biogenic reefs using airborne optical imagery and neural network: the honeycomb worm (Sabellaria alveolata) case study. Int J Remote Sens 39(17):5660–5675
7. Collin A, Durozier T, Hervouet G, Costa S (2021) Détecter le processus de dépoldérisation par l'observation satellite à très haute résolution temporelle. 16th SAGEO, May 2021, pp 250–252
8. Collin A, James D, Mury A, Letard M, Guillot B (2021) Predicting the infrared UAV imagery over the coast. ISPRS-International Archives of the Photogrammetry. Remote Sens Spat Inf Sci 43:149–156
9. Collin A, Letard M, Andel M, Sharma S (2021) Object-based mangrove mapping using sub-meter superspectral WorldView-3 imagery and deep convolutional neural network. IEEE International Geoscience and Remote Sensing Symposium, July 2021, pp 7362–7365
10. Collin A, Boymond C, Dubois S (submitted) How the largest coastal animal reef facilitates one of the broadest vegetal meadow in Europe: the mineral hyphen. Int Temp Reefs Symp
11. Deegan LA, Johnson DS, Warren RS, Peterson BJ, Fleeger JW, Fagherazzi S, Wollheim WM (2012) Coastal eutrophication as a driver of salt marsh loss. Nature 490(7420):388–392
12. Domínguez-Beisiegel M, Castañeda C, Mougenot B, Herrero J (2016) Analysis and mapping of the spectral characteristics of fractional green cover in saline wetlands (NE Spain) using field and remote sensing data. Remote Sens 8(7):590
13. Donatelli C, Ganju NK, Zhang X, Fagherazzi S, Leonardi N (2018) Salt marsh loss affects tides and the sediment budget in shallow bays. J Geophys Res Earth 123(10):2647–2662
14. Godet L, Pourinet L, Joyeux E, Verger F (2015) Dynamique spatiale et usage des schorres de l'Anse de l'Aiguillon de 1705 à nos jours. Enjeux de conservation d'un patrimoine naturel littoral marin. Cyber Eur J Geogr
15. Gu J, Luo M, Zhang X, Christakos G, Agusti S, Duarte CM, Wu J (2018) Losses of salt marsh in China: trends, threats and management. Estuar Coast Shelf Sci 214:98–109
16. Hardisky MA, Daiber FC, Roman CT, Klemas V (1984) Remote sensing of biomass and annual net aerial primary productivity of a salt marsh. Remote Sens Environ 16(2):91–106
17. Heermann PD, Khazenie N (1992) Classification of multispectral remote sensing data using a back-propagation neural network. IEEE Trans Geosci Remote Sens 30(1):81–88
18. James D, Collin A, Mury A, Letard M, Guillot B (2021) UAV multispectral optical contribution to coastal 3D modelling. IEEE Int Geosci Remote Sens Symp IGARSS 2021:7951–7954
19. Kirwan ML, Guntenspergen GR (2012) Feedbacks between inundation, root production, and shoot growth in a rapidly submerging brackish marsh. J Ecol 100(3):764–770
20. Kirwan ML, Megonigal JP (2013) Tidal wetland stability in the face of human impacts and sea-level rise. Nature 504(7478):53–60
21. Le Quilleuc A, Collin A, Jasinski M, Devillers R (2022) Very high-resolution satellite derived bathymetry and habitat mapping using Pleaides-1 and ICESat-2. Remote Sens 14(1):133

22. Letard M, Collin A, James D, Mury A (2021) Trajectoire socio-écologique d'une commune de la Baie du Mont-Saint-Michel entre 1947 et 2014. 16th SAGEO, May 2021, pp 63–74
23. Loconsole D, Cristiano G, De Lucia B (2019) Glassworts: from wild salt marsh species to sustainable edible crops. Agriculture 9(1):14
24. Lopes CL, Mendes R, Caçador I, Dias JM (2020) Assessing salt marsh extent and condition changes with 35 years of Landsat imagery: Tagus Estuary case study. Remote Sens Environ 247:111939
25. Mury A, Collin A, James D (2019) Morpho–sedimentary monitoring in a coastal area, from 1D to 2.5 D, using airborne drone imagery. Drones 3(3):62
26. Mury A, Collin A, Houet T, Alvarez-Vanhard E, James D (2020) Using multispectral drone imagery for spatially explicit modeling of wave attenuation through a salt marsh meadow. Drones 4(2):25
27. Mury A, Collin A, Etienne S, Jeanson M (2020) Wave attenuation service by intertidal coastal ecogeosystems in the Bay of Mont-Saint-Michel, France: review and meta-analysis. In: Estuaries and coastal zones in times of global change, pp 555–572
28. Mury A, Collin A, Jeanson M, James D, Gloria H, Pastol Y, Etienne S (2020) Mapping nature-based marine flooding risk using VHR wave, airborne LiDAR and satellite imagery: the case study of the Dol Marsh (Bay of Mont-Saint-Michel, France). J Coast Res 95(SI):743–747
29. Planet Team (2017) Planet application program interface: in space for life on Earth. San Francisco
30. Proença B, Frappart F, Lubac B, Marieu V, Ygorra B, Bombrun L, Sottolichio A (2019) Potential of high-resolution Pléiades imagery to monitor salt marsh evolution after Spartina invasion. Remote Sens 11(8):968
31. Rendón OR, Garbutt A, Skov M, Möller I, Alexander M, Ballinger R, Beaumont N (2019) A framework linking ecosystem services and human well-being: saltmarsh as a case study. People Nat 1(4):486–496
32. Rouse JW, Haas RH, Schell JA, Deering DW (1974) Monitoring vegetation systems in the Great Plains with ERTS. NASA Spec Publ 351(1974):309
33. Russell RJ, Morgan JP (1952) Trafficability and navigability of delta-type coasts. Trafficability and navigability of Louisiana coastal marshes. Photo-interpretation keys of selected coastal marshland features
34. Saintilan N, Rogers K, McKee KL (2019) The shifting saltmarsh-mangrove ecotone in Australasia and the Americas. In: Coastal wetlands, pp 915–945
35. Stroud LM, Cooper AW (1969) Color-infrared aerial photographic interpretation and net primary productivity of a regularly flooded North Carolina salt marsh. Water Resources Research Institute of the University of North Carolina
36. Sun C, Li J, Cao L, Liu Y, Jin S, Zhao B (2020) Evaluation of vegetation index-based curve fitting models for accurate classification of salt marsh vegetation using Sentinel-2 time-series. Sensors 20(19):5551
37. Taramelli A, Valentini E, Cornacchia L, Monbaliu J, Sabbe K (2018) Indications of dynamic effects on scaling relationships between channel sinuosity and vegetation patch size across a salt marsh platform. J Geophys Res Earth 123(10):2714–2731
38. Torrecilla E, Piera J, Pons S, Aymerich IF, Vilamala A, Arcos JL, Plaza E (2010) Mapping marine phytoplankton assemblages from a hyperspectral and artificial intelligence perspective. Ocean's 10 IEEE, Sydney, Australia, May 2010, pp 1–7
39. Valéry L, Radureau A, Lefeuvre JC (2017) Spread of the native grass Elymus athericus in salt marshes of Mont-Saint-Michel Bay as an unusual case of coastal eutrophication. J Coast Conserv 21(3):421–433

Contribution of Near- and Mid-Infrared Wavebands to Mapping Fine-Scale Coastal Ecogeomorphological Features

Antoine Mury, Antoine Collin, Dorothée James, and Mathilde Letard

Abstract Coastal ecogeomorphological features support remarkable biodiversity and provide a wide variety of ecosystem services: cultural services (recreation, tourism facilities), provisioning services (agricultural production, pastoralism), and regulating services, including carbon sequestration and natural protection against coastal erosion and marine flooding. Therefore, mapping these coastal features with very high spatial resolution is a major challenge to their preservation and to face the challenges of global change. In this study, the contribution of the near-infrared (NIR) and mid-infrared (MIR) bands from multispectral drone and super-spectral (SS) WorldView-3 (WV-3) satellite images was used to map coastal ecogeomorphological features using two supervised classification algorithms: maximum likelihood (ML) and support vector machine (SVM). Various combinations of spectral bands, visible + NIR and visible + MIR, evaluated through the overall accuracy (OA) scores, for the classification of ecogeomorphological features revealed the significant contribution of the NIR and MIR bands to the mapping of coastal features. The addition of the NIR bands to the RGB band combination significantly increased the OA scores of the classifications (by +4.99% and +6.54%, with the ML and SVM algorithms, respectively). The addition of MIR bands to the combination of these bands provides classifications with even higher OAs (up to 99.1% and 98.4%), demonstrating the relevance of MIR bands for the mapping of coastal ecogeomorphological features.

A. Mury (✉) · D. James · M. Letard
Ecole Pratique des Hautes Etudes – PSL Université, Dinard, France
e-mail: antoine.mury@etu.ephe.psl.eu; dorothee.james@ephe.psl.eu; mathilde.letard@ephe.psl.eu

A. Collin
Ecole Pratique des Hautes Etudes – PSL Université, Dinard, France

Laboratoire d'excellence CORAIL, Perpignan, France
e-mail: antoine.collin@ephe.psl.eu

S. Niculescu (ed.), *European Spatial Data for Coastal and Marine Remote Sensing*, https://doi.org/10.1007/978-3-031-16213-8_3

Keywords Classification · Coastal features · Mapping · Maximum likelihood · Super-spectral satellite imagery · Support vector machine

1 Introduction

The exceptional contribution of ecosystem services provided by the coastal systems of the Bay of Mont-Saint-Michel (BMSM) has been well documented [14, 17, 22]. The ecogeomorphological characteristics of these wetlands, such as their floristic compositions, their root systems, and their ability to reduce the waves' height through their densities, are critical to ensure the resilience of the landscape against deleterious climate change impacts and also protection of indigenous societies against coastal hazards [13, 18–20]. In addition, coastal wetlands support a wide range of ecological infrastructure and biodiversity assets, recognized through various national and international treaties. In Europe, this includes, for example, the European "birds" (council directive 2009/147/EC) and "habitats" (council directive 92/43/EEC) directives. In doing so, coastal wetlands were selected for inclusion in protected areas, such as Natura 2000, natural zones of ecological interest, and the Ramsar Convention on wetlands of international importance. Therefore, the identification and mapping of ecogeomorphological features are carried out as preliminary steps toward the management and preservation of natural environments.

Recognized as a UN Educational, Scientific, and Cultural Organization World Heritage Site since 1979 (whc.unesco.org), the BMSM constitutes an important biodiversity reserve and a real regional wealth. In fact, it offers local communities a wide range of benefits in terms of ecosystem services.

Previously, two coastal ecogeomorphological features of the BMSM have been mapped, including the biogenic reefs of *Sabellaria alveolata* [8, 9] and salt marshes [19], both using remote sensing methods based on Light Detection and Ranging (LiDAR) and satellite or unmanned aerial system (UAS) data. But no study to date has focused on mapping the complex set of ecomorphological systems at the land–sea interface. However, a global understanding of spatial dynamics over time is necessary given the existing interactions between the ecogeomorphological entities, as illustrated by the notion of "windows of opportunity" [3, 12].

In this study, the contribution of the novel bands from the spaceborne, super-spectral WorldView-3 (WV-3) satellite imagery to the visible red-green-blue (RGB) bands combination is evaluated for the very high-resolution (VHR) mapping of ecogeomorphological features, through two conventional classification algorithms, ML and SVM, in order to identify the most useful spectral bands combination for mapping, regardless of the algorithm used.

2 Methodology

2.1 Study Site

The BMSM (France) is located in the Brittany–Normandy Gulf between the Cotentin Peninsula and the northern coast of Brittany called the "Emerald Coast." This bay, subjected to a megatidal regime, belongs to the top six areas that host the world's highest tide [1]. The bay stretches from the Dol Marsh polder by the Duchess Anne's dike in the south to a virtual line between the point of Grouin and the point of Granville in the north. In the south, extensive salt marshes of approximately 42 km^2 [24] offer coastal protection by acting as natural barriers against extreme tidal action. In addition to these salt marshes, the bay also houses original geomorphological features called "shelly ridges," which were formed in part by thanatocenosis from calcareous bivalve shells mixed with siliciclastic sediments. Salt marshes and ridges play an important ecological role by sheltering protected species such as *Charadrius alexandrines* [LC] (www.iucnredlist.org) and offering natural protection. This study area is a 0.1 km^2 area within the bay, which contains coastal salt marshes and ridges, such as wetlands and ecogeomorphological features that deserve to be mapped more accurately (Fig. 1) for a full understanding of their shapes and dynamics. The mapping of these different features, whether plant (salt marshes) or sedimentary (mudflats, as well as the different stages of evolution of the shelly ridges), constitutes a major action of bringing awareness to their preservation.

2.2 Imagery Data

The areal extent of the ecogeomorphological features of the landscape, including shelly ridges and mudflats, as well as wetlands, were mapped using a single Worldview–3 (WV-3) image, acquired by the European Spatial Agency (ESA) in February 10, 2019. The WV-3 optical sensor offers one panchromatic band and 16 multispectral bands (Fig. 2), of which eight range from visible to NIR:

Band	Wavelength (nm)
Coastal [C]	400–450
Blue [B]	450–510
Green [G]	510–580
Yellow [Y]	585–625
Red [R]	630–690
Red edge [RE]	705–745
Near-infrared 1 [NIR1]	770–895
Near-infrared 2 [NIR2]	860–1040

eight from MIR to shortwave infrared [SWIR]:

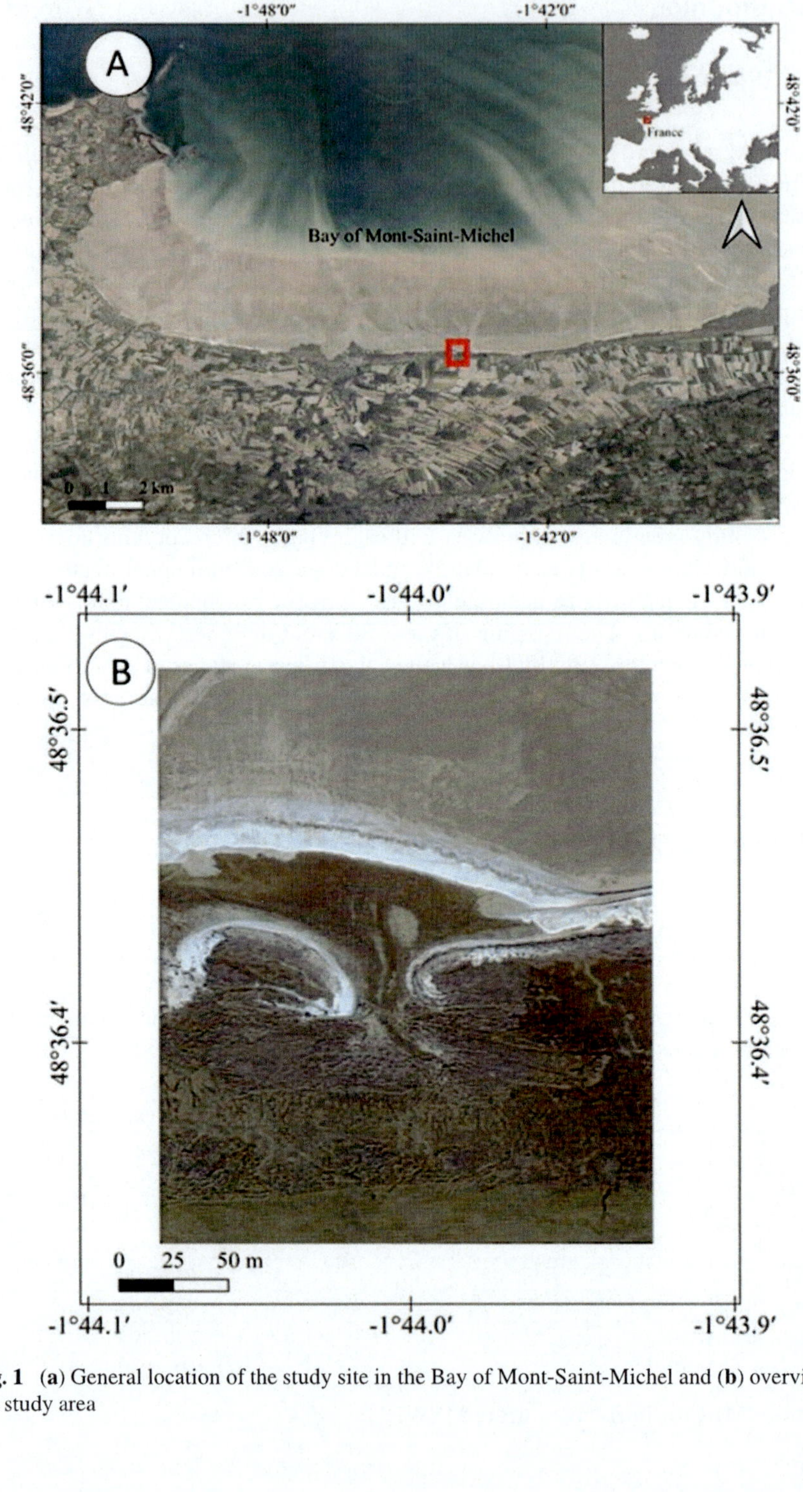

Fig. 1 (**a**) General location of the study site in the Bay of Mont-Saint-Michel and (**b**) overview of the study area

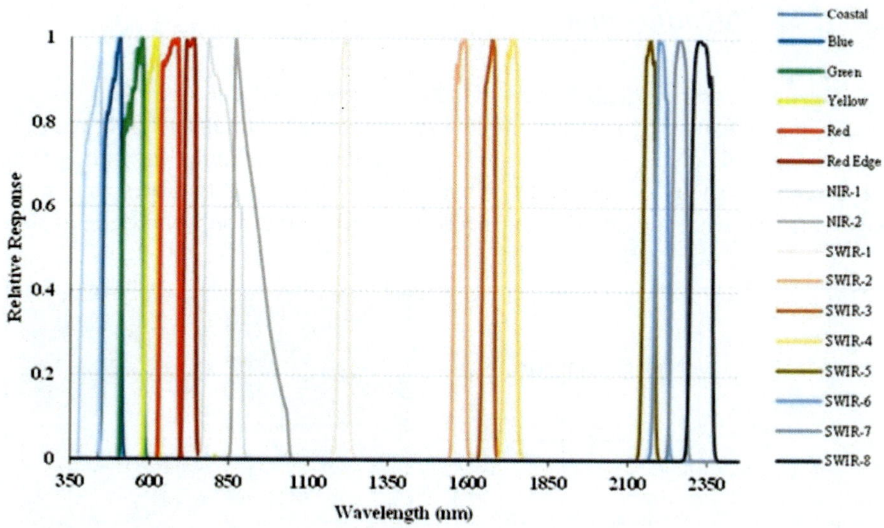

Fig. 2 Line plot of the relative response of the 16 spectral bands of the WorldView-3 space-borne sensor

Band	Wavelength (nm)
Shortwave infrared 1 [SWIR1]	1195–1225
Shortwave infrared 2 [SWIR2]	1550–1590
Shortwave infrared 3 [SWIR3]	1640–1680
Shortwave infrared 4 [SWIR4]	1710–1750
Shortwave infrared 5 [SWIR5]	2145–2185
Shortwave infrared 6 [SWIR6]	2185–2225
Shortwave infrared 7 [SWIR7]	2235–2285
Shortwave infrared 8 [SWIR8]	2295–2365

at 0.3, 1.2, and 3.7 m pixel size, respectively (www.digitalglobe.com).

Super-spectral (SS) images underwent three pre-processing steps, and the influence of the atmosphere was corrected (top of atmosphere, TOA) for each image. The radiometric correction of the SS images consists in converting the numerical radiance values into TOA reflectance values. The reflectance is calculated according to the radiance/irradiance (solar) ratio. The images were also geometrically corrected. The spatial resolution of the imagery was then resampled to that of the panchromatic band (0.3 m) using the Gram–Schmidt pan-sharpening procedure. An ultrahigh resolution (<0.03 m) UAS georeferenced orthophoto mosaic (achieved using Pix4d mapper software) with bands centered in the red (660 ± 30 nm), green (545 ± 35 nm), and blue (480 ± 30 nm) ranges of the electromagnetic spectrum was used to validate the classification of coastal features that were mapped using WV-3.

2.3 Class Identification

Seven ecogeomorphological classes, which can be divided into two main categories, (i) plant features (including three sub-categories, natural and pastured salt marshes and tidemark plant debris) and (ii) sedimentary features (including four sub-categories, shelly ridges, shelly outwashes, in-process shelly ridges, and mudflats) (Fig. 3), were identified using georeferenced ground-truth photographs and

Fig. 3 Photographs showing the seven coastal ecogeomorphological sub-classes and two main classes used in this study

the UAS imagery. A series of 2000 regions of interest, single pixels from a single WV-3 image for each class, was manually selected in the form of spectrally homogeneous polygons (Fig. 4). This data set was split into 50% calibration and 50% validation pixels for each class.

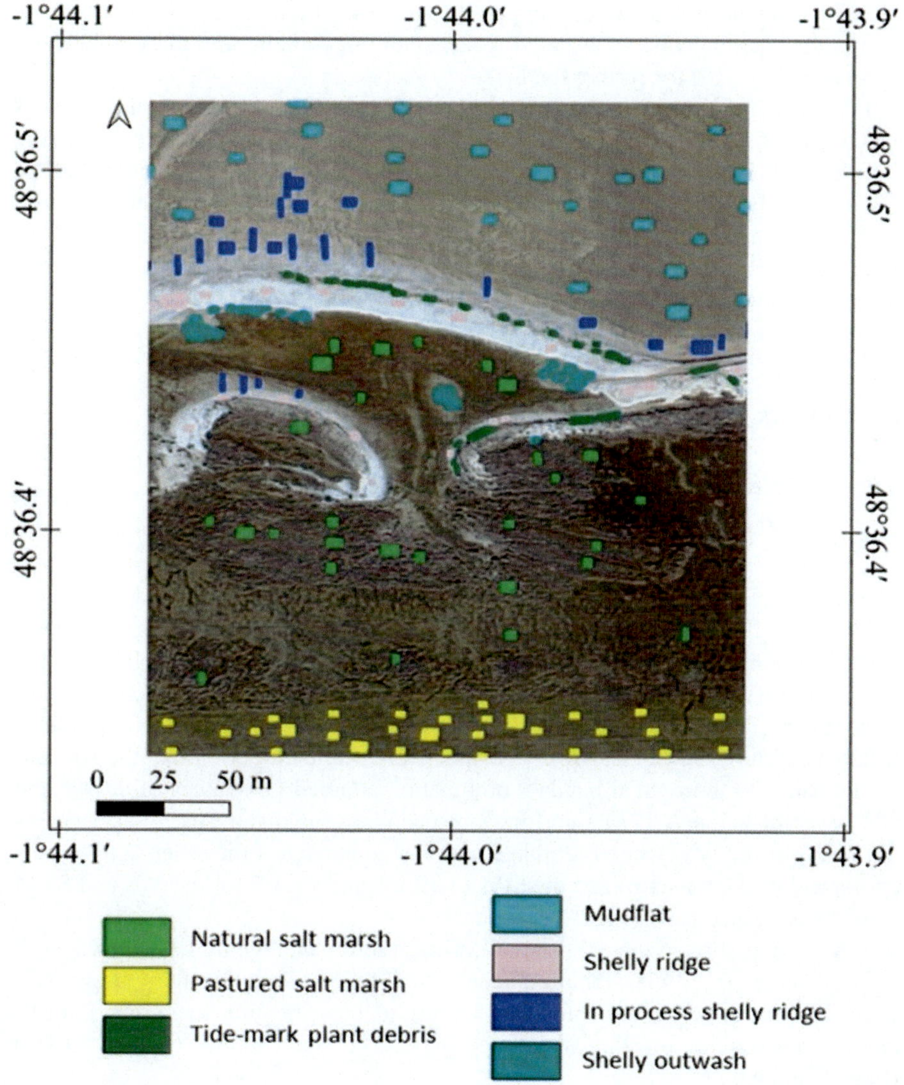

Fig. 4 Location map of the selected pixels as regions of interest for each of the seven fine-scale classes related to coastal ecogeomorphological features

2.4 Pixel-Based Supervised Classification

Two supervised classification algorithms based on statistical learning theory were compared: the common and fast ML and the complex-solving but time-consuming SVM. ML is a probabilistic method, assuming that the statistics for each class in each band are normally distributed, and which calculates the probability that a given pixel belongs to a specific class. Each pixel is assigned to the highest probability class.

SVM is a non-probabilistic, non-parametric binary multiclassifier that separates the classes with a decision surface (hyperplane), maximizing the margin between the classes (see [15] for further details).

Five classification scenarios were compared based on a selection of the WV-3 spectral bands (i) using the RGB spectral bands only, (ii) RGB and NIR, (iii) visible to SWIR4, (iv) from visible to NIR plus SWIR5 to SWIR8, and (v) all of the WV-3 bands. The best model was firstly considered where the classification results attained the highest overall accuracy (OA) and secondly when also considering the maximum producer's accuracy (PA) for each ecogeomorphological class.

3 Results and Discussion

3.1 Identification of the Best Model Based on Overall Accuracy

The classifications of the visible RGB bands presented an OA of 89% and 87% for the ML and SVM algorithms, respectively. The addition of the other visible bands (coastal [C] and the yellow [Y] bands) and optical bands (NIR1 and NIR2) of the WV-3 imagery for the classification increased the OA scores to 94% and 93.6%, respectively. The addition of the first four SWIR bands to the eight previous bands, resulted in OAs of 98.2% and 98.3%, respectively, while the addition of the last four SWIR bands to the optical bands' combination attained OA values of 97.9% and 98.2% for the ML and SVM algorithms, respectively. Finally, the use of all 16 spectral bands of the WV-3 satellite imagery, for the mapping of the seven ecogeomorphological classes, maximized the OAs to 99.1% and 98.4% for the ML and SVM algorithms, respectively (Figs. 5 and 6).

OAs obtained from the ML and SVM differed on average by only 0.68% with the OA values in all classification scenarios (Fig. 5). Based on the comparative OAs reached by both algorithms for the various classification scenarios, the ML algorithm using all of the WV-3 bands was the best model (OA = 99.1%) (Figs. 5 and 6).

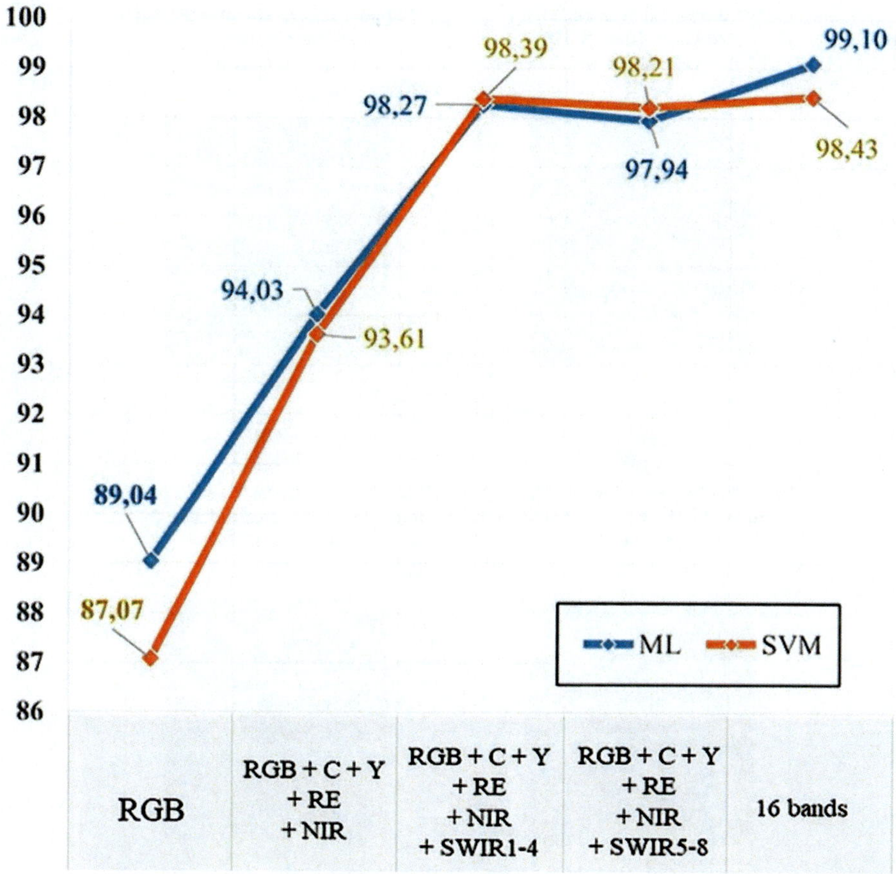

Fig. 5 Line plot of the overall accuracies (OAs) obtained for each of the five classification scenarios (1) Red-Green-Blue [RGB], (2) RGB-Coastal [C]-Yellow [Y]-Near Infrared [NIR], (3) RGB-C-Y-NIR-Shortwave Infrared [SWIR] 1–4, (4) RGB-C-Y-NIR-SWIR5-8, and (5) all 16 bands and two classification algorithms. (Maximum likelihood, ML, and support vector machine, SVM, algorithms)

3.2 *Class-Level Accuracy*

3.2.1 Best Model for Classification of the Plant Features

In terms of classifications of the plant features, the classification results of natural salt marshes using the RGB bands only resulted in PA values of 96.5% and 94.1% for the ML and SVM algorithms, respectively. The PAs attained 99.7% and 98.9%, respectively, with the addition of the other visible bands and NIR bands and 99.8%

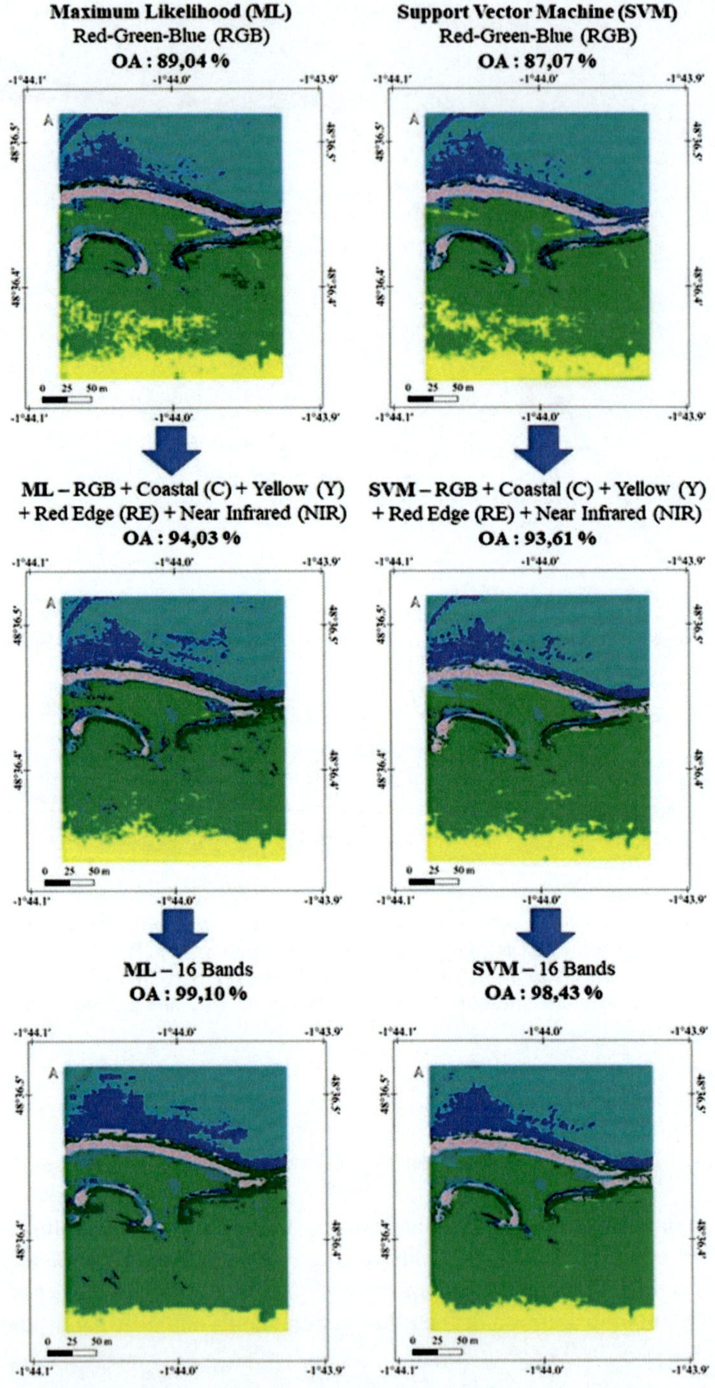

Fig. 6 Maps resulting from the various classification scenarios and two classification algorithms used to map the coastal ecogeomorphological features of the Bay of Mont-Saint-Michel

and 100% when the SWIR1 to SWIR4 bands were added to this previous combination. The PA values were 99.7% and 99.2%, respectively, when SWIR5 to SWIR8 were included with the RGB-C-Y-RE-NIR band combination classification scenario and maximized to a PA of 99.9% and 100%, respectively, when all 16 bands of the WV-3 imagery were used (Fig. 7).

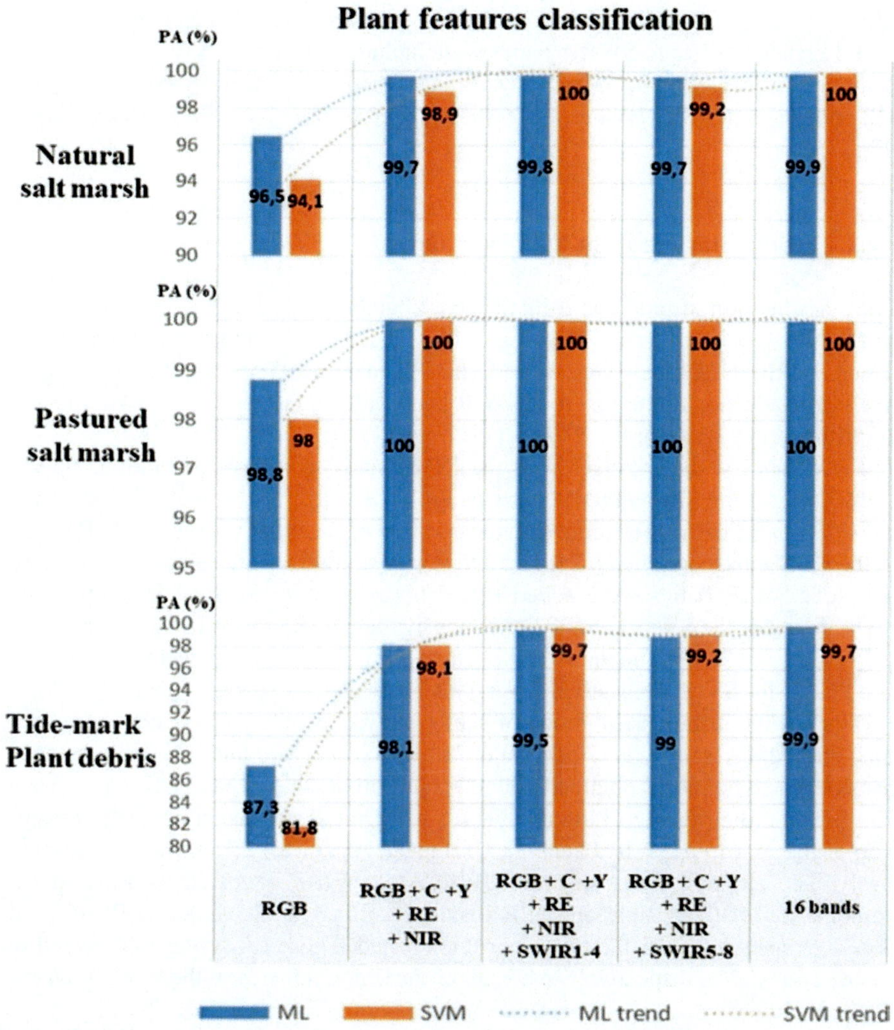

Fig. 7 Bar graphs of the producer's accuracy obtained for plant feature classifications, using maximum likelihood (ML) and support vector machine (SVM) algorithms and several bands' combinations (1) Red-Green-Blue [RGB], (2) RGB-Coastal [C]-Yellow [Y]-Near Infrared [NIR], (3) RGB-C-Y-NIR-Shortwave Infrared [SWIR] 1–4, (4) RGB-C-Y-NIR-SWIR5-8, and (5) all 16 bands

PA values obtained for pastured salt marshes increased from 98.8% for the ML algorithm to 98% for the SVM classifications using only the RGB bands, to 100% for both classification algorithms when including the NIR bands (Fig. 7).

The classification accuracy for the tidemark plant debris class followed the same trend as the salt marshes: from PAs attaining 87.3% with ML and 81.8% with SVM in the RGB classification scenario to 98.1% for both algorithms, with the addition of the NIR bands. The PA values increased to 99.5% and 99.7% for the ML and SVM classification scenarios, respectively, when the SWIR1 to SWIR4 bands were added to the RGB-NIR band classification scenario. The PA values decreased slightly when SWIR5 to SWIR8 bands were included, instead of SWIR1 to SWIR4, with 99% for ML and 99.2% for SVM. Finally, the PA was maximized at 99.9% for ML and 99.7% for SVM, when all 16 spectral bands of the WV-3 images were used (Fig. 7).

3.2.2 Sedimentary Feature Classifications

The classification of mudflats from the RGB bands resulted in PA scores of 92.4% and 92.9% for the ML and SVM algorithms, respectively. When the other visible bands and the NIR bands were added, this PA reached 95.5% and 98.4%, respectively. The addition of the first four SWIR bands allowed to reach 95.7% for the ML algorithm and 99.6% for the SVM algorithm (Fig. 8).

PA values for shelly ridges with the RGB bands' combination attained 93.9% and 93.5% for the ML and SVM classification scenarios, respectively. These scores were 95.3% for both algorithms with the addition of the other visible and NIR bands to the RGB bands and then maximized to 96.3% and 96.5%, respectively, when the first four SWIR bands were added to the classification scenario. Thereafter, the replacement of the SWIR1 to SWIR4 bands by the SWIR5 to SWIR8 bands resulted in PA scores of 96.1% and 96.5%, respectively. Finally, PA scores were 95% for ML and 96.4% for SVM, respectively, using all 16 of WV-3's spectral bands (Fig. 8).

The results of the classification of shelly outwashes, the RGB bands' combination offered PA scores of 74.1% and 71.6% for the ML and SVM algorithms, respectively. The scores obtained with the addition of the C, Y, and NIR bands were 79.9% using the ML algorithm and 78.1% using the SVM algorithm. The integration of the first four SWIR bands then achieved PAs of 99.2% and 97.6%, respectively. The classification using the SWIR5 to SWIR8 bands in addition to the combination of the visible and NIR bands, resulting in PA values of 98.4% and 96.2%, respectively, for the two algorithms tested. These PA scores maximized to 99.6% and 97.6%, respectively, when all of the spectral bands of the WV-3 imagery are used (Fig. 8).

Finally, the PA for in-process shelly ridges attained 80.3% for ML and 77.6% for SVM when using the RGB bands only and then 89.7% and 86.5%, respectively, for ML and SVM, with the addition of the other visible and NIR bands. The PAs of each algorithm reached 93.5% and 95.3%, respectively, when the SWIR1 to SWIR4 bands were added to the classification scenario. The PAs then attained 94.5% and

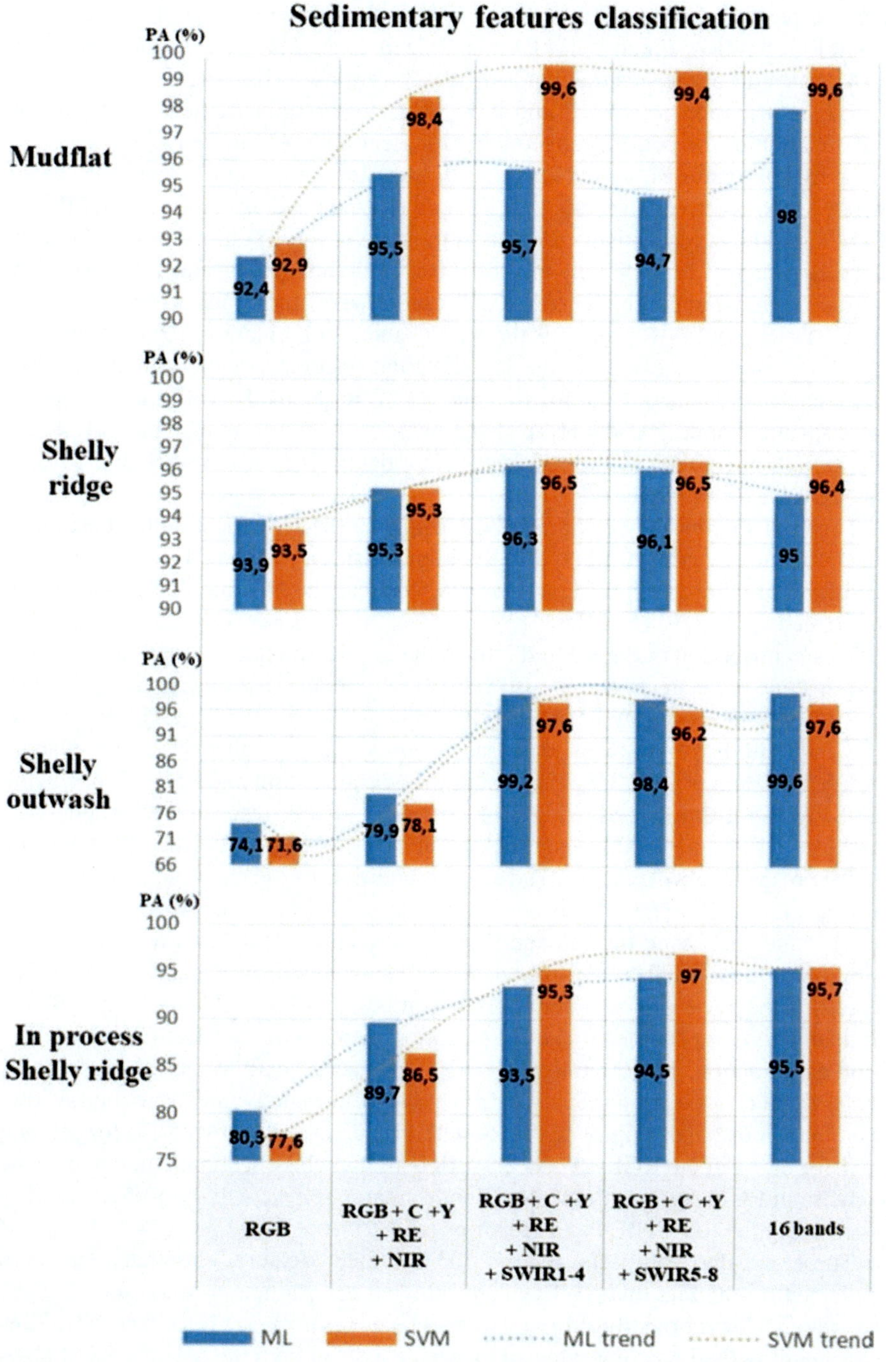

Fig. 8 Bar graphs of the producer's accuracy obtained for sedimentary feature classifications using maximum likelihood (ML) and support vector machine (SVM) algorithms and several bands' combinations (1) Red-Green-Blue [RGB], (2) RGB-Coastal [C]-Yellow [Y]-Near Infrared [NIR], (3) RGB-C-Y-NIR-Shortwave Infrared [SWIR] 1–4, (4) RGB-C-Y-NIR-SWIR5-8, and (5) all 16 bands

97%, respectively, when the SWIR5 to SWIR8 bands were used rather than bands SWIR1 to SWIR4. Finally, the PA maximized to 95.5% for ML and 95.7% for SVM, when all of the spectral bands of WV-3 were used (Fig. 8).

The contribution of the RE and NIR bands to the improvement of the overall classification of the coastal ecogeomorphological features, as well as the accuracy for individual features, was clear. Indeed, the addition of the RE and NIR bands increases the producer's accuracy by 5% and 6.5% on average while using ML and SVM algorithms, respectively. In fact, in the case of the classifications of the plant features category (natural salt marshes, pastured salt marsh, and tidemark plant debris), the most important increases in PA score were attributable to the addition of the C, Y, RE, and NIR bands to the RGB bands' combination, regardless of the algorithm used. For example, the classification accuracy of the salt marsh class increased by 3.2% and 4.8% for ML and SVM, respectively, with the addition of these spectral bands. PA scores are also grown for pastured salt marshes (+1.2% for ML and 2% for SVM) and for tidemark plant debris (+10.8% for ML and +16.3% for ML) (Fig. 9).

The contribution of the RE and NIR bands to the classification of these plant coastal features can be attributed to the high sensitivity of the RE to a medium to high level of chlorophyll content of plants, whereas the NIR highlights the vegetation structure [6, 7, 9, 10, 20].

This contribution of the RE and NIR to the improvement of coastal feature mapping at a fine scale, in general, is also valid to sedimentary features, although the OA and PA only increased by between 1.4% and 9.4% (Fig. 9). The addition of C, Y, RE, and NIR bands showed increases in PA values for all classes: mudflat by between 3% and 6% (+3.1% when using the ML algorithm and +5.5% when using the SVM algorithm), shelly ridges (+1.4% and +1.8%, respectively), shelly outwashes (+5.8% and +6.5%, respectively), and in-process shelly ridges (+9.4% and +8.9%, respectively) (Fig. 9). However, in addition to the improvement resulting from using the RE and NIR bands in the classification scenarios, the use of the SWIR bands improved the OA and PAs of sedimentary features. For example, the addition of the SWIR1 to SWIR4 bands to the visible-RE-NIR-band combinations increased the PA score of mudflat classification by 0.2% (ML) and 1.2% (SVM) compared to those attained in the classification scenario of the RGB-C-Y-RE-NIR--band combination. The same trends were visible with shelly ridges and in-process shelly ridge classification scenarios. In the case of the shelly outwashes class, however, the SWIR bands improved the PA with an increase in PA by 25.1% for ML and 26% for SVM in the RGB-RE-NIR-SWIR1-4-band classification, compared to the increase in PA of 5.8% with the ML algorithm and +6.5% with the SVM algorithm when using the RGB-NIR band scenarios (Fig. 9).

These satisfactory performances for sediment feature classifications were explained by the fact that SWIR bands allowed determination of several physical and chemical soil properties through reflectance spectroscopy [2, 4, 5, 16, 21, 23].

Regarding the PA scores attained by each class, for both the ML and SVM algorithms, all 16 spectral bands of the WV-3 imagery were used, and four of them presented values lower than the OA: mudflats (98% with the ML algorithm), shelly

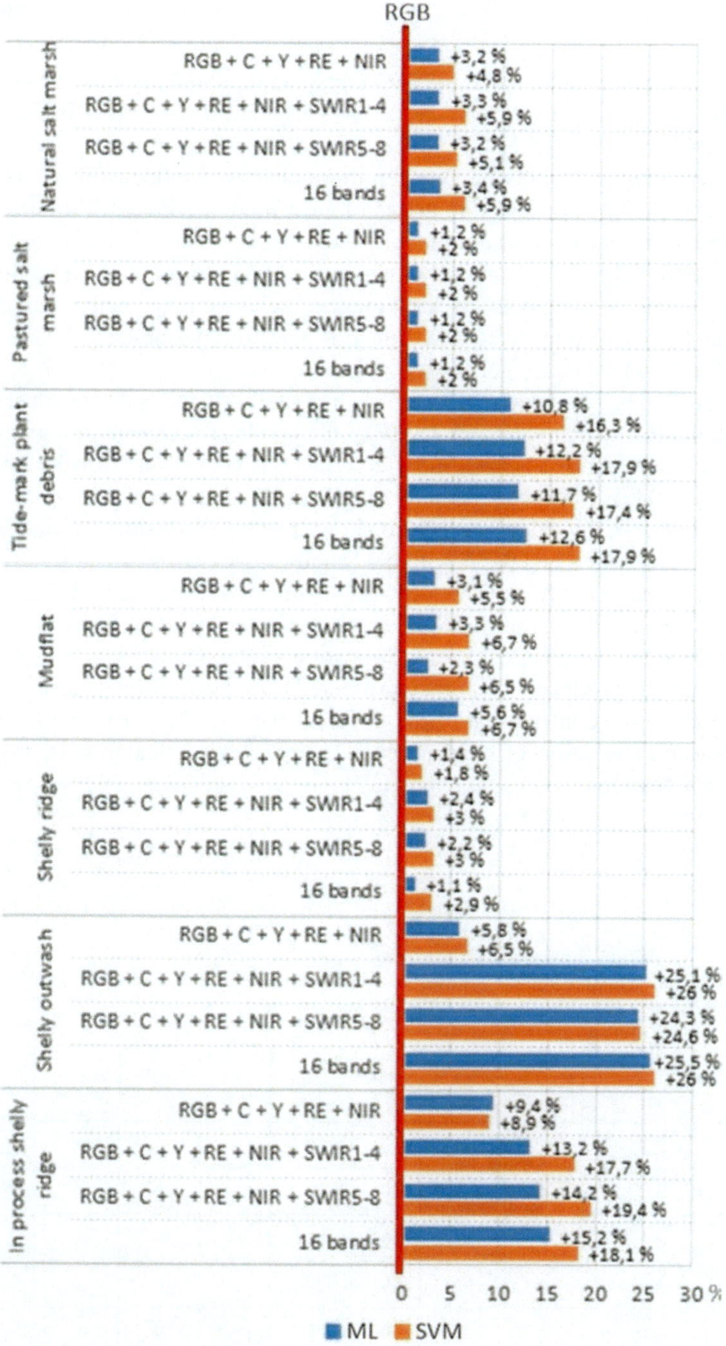

Fig. 9 Bar plots showing the increase in the producer's accuracy for each individual coastal ecogeomorphic feature class, using one of five classification scenarios. Two algorithms were used in the scenarios, namely, the maximum likelihood (ML) and support vector machine (SVM), while several band combinations were included (1) Red-Green-Blue [RGB], (2) RGB-Coastal [C]-Yellow [Y]-Near Infrared [NIR], (3) RGB-C-Y-NIR-Shortwave Infrared [SWIR] 1-4, (4) RGB-C-Y-NIR-SWIR5-8, and (5) all 16 bands

ridges (95% for ML and 96.4% for SVM, respectively), shelly outwashes (97.6% for SVM), and in-process shelly ridges (95.5% and 95.7% for ML and SVM, respectively).

Recurrent spectral confusion between these classes is reflected in the relatively low PA scores of the confusion matrices of each model, with Table 1 showing the confusion matrix of the optimum model. Here it is clear that all pixels belonging to the plant feature classes are correctly assigned, while in the case of the sedimentary features, 98.8% of the pixels of the "mudflat" class are correctly assigned, the remaining 1.2% being incorrectly assigned to the "shelly outwash" (1%) and "in-process shelly ridge" (0.2%) classes.

This can also be observed in the case of the "shelly ridge" and "in-process shelly ridge" classes for which 98% and 96.9% of the pixels are correctly assigned. In the case of the "shelly ridge" class, there is a spectral confusion with the "tidemark plant debris" (0.8%), "in-process shelly ridge" (0.7%), and "shelly outwash" (0.5%). The "in-process shelly ridge" class also show spectral confusion of 2.6% with the "shelly outwash" class and 0.5% with the "shelly ridge" class.

Existing spectral confusions between the several classes found their explanations in the physical similarities of these sedimentary features (mix of fine sediments and coarse shell debris, in different proportions for each class).

The results of producer's accuracy presented previously (Figs. 6 and 7) showed some trends on the contribution of the different parts of the electromagnetic spectrum to the classification of coastal ecogeomorphological features, be it plant or sedimentary features.

Table 1 Confusion matrix of the optimum model (maximum likelihood using the 16 WorldView-3 spectral bands)

Class	Natural salt marsh	Pastured salt marsh	Tide mark plant debris	Mudflat	Shelly ridge	Shelly outwash	In process shelly ridge
Natural salt marsh	100	0	0	0	0	0	0
Pastured salt marsh	0	100	0	0	0	0	0
Tide mark plant debris	0	0	100	0	0.8	0	0
Mudflat	0	0	0	98.80	0	0	0
Shelly ridge	0	0	0	0	98.0	0	0.5
Shelly outwash	0	0	0	1.0	0.5	100	2.6
In process shelly ridge	0	0	0	0.2	0.7	0	96.9
Total	100	100	100	100	100	100	100

4 Conclusion

The use of the WorldView-3 satellite imagery, in particular the red-edge, NIR, and SWIR bands, has offered improved mapping of coastal ecomorphological features in the Bay of Mont-Saint-Michel, France. The ML algorithm attained the best classification results using all the spectral bands (maximum overall accuracy of 99.1%), while the support vector machine algorithm resulted in 98.4%.

The different combinations of bands experimented shed light on the interest of the different parts of the electromagnetic spectrum for the classification of coastal features, both plant and sedimentary. Thus, it appears, in accordance with the literature, that the spectral bands in the near-infrared range are particularly relevant for the mapping of different vegetation classes (natural salt marsh, pastured salt marsh, and tidemark plant debris) [11]. In effect, the addition of these spectral bands, particularly the red edge and NIR, to the three traditional bands of the visible spectrum, red, green, and blue, increases the individual producer's accuracies of these plant feature classes by between 1.2% and 16.3% (Fig. 9).

For the sedimentary features, the addition of the RE, NIR, and SWIR improved the OA and individual PA of the classification.

These increases in accuracy could be explained by the spectral properties of SWIR bands, which allow us to study the soil characteristics through reflectance spectroscopy.

Finally, from the different results obtained in this study, it emerges that the joint use of the NIR and SWIR spectral bands in addition to the visible spectrum bands offers the best classification results. The contribution of each of these ranges of the infrared spectrum is particularly significant and their complementarity is undeniable for the study of natural coastal sites with both plant and sedimentary systems.

This work, carried out on a geographically restricted area, presents a generic methodology and results that can be applied to the whole of the Bay of Mont-Saint-Michel and to other geographical sectors, on the condition that the phase of acquiring ground truths, in particular their spatial distribution, is adapted to the different study sites.

Acknowledgments We deeply acknowledge the European Space Agency for their support in WorldView-3 satellite imagery acquisition.

References

1. Archer AW (2013) World's highest tides: hypertidal coastal systems in North America, South America and Europe. Sediment Geol 1–25:284–285
2. Asner GP, Lobell DB (2000) A biogeophysical approach for automated SWIR unmixing of soils and vegetation. Remote Sens Environ 74(1):99–112
3. Balke T, Bouma TJ, Horstman EM, Webb EL, Erftemeijer PLA, Herman PMJ (2011) Windows of opportunity: thresholds to mangrove seedling establishment on tidal flats. Mar Ecol Prog Ser 440:1–9

4. Ben-Dor E, Chabrillat S, Dematte JAM, Taylor GR, Hill J, Whiting ML, Sommer S (2009) Using imaging spectroscopy to study soil properties. Remote Sens Environ 113:S38–S55
5. Brosinsky A (2015) Spectral fingerprinting: the potential of VNIR-SWIR spectral characteristics for tracing suspended sediment sources. Thesis, University of Potsdam, 127 p
6. Carter GA, Knapp AK (2001) Leaf optical properties in higher plants: linking spectral characteristics to stress and chlorophyll concentration. Am J Bot 88:677–684
7. Collin A, Lambert N, Etienne S (2018) Satellite-based salt marsh elevation, vegetation height, and species composition mapping using the superspectral Worldview-3 imagery. Int J Remote Sens 39(17):5619–5637
8. Collin A, Dubois S, James D, Houet T (2019) Improving intertidal reef mapping using UAV surface, red edge, and near-infrared data. Drones 3:67
9. Collin AM, Andel M, James D, Claudet J (2019) The superspectral/hyperspatial WorldView-3 as the link between spaceborne hyperspectral and airborne hyperspatial sensors: the case study of the complex tropical coast. Int Arch Photogramm Remote Sens Spat Inf Sci XLII-2/W13:1849–1854
10. Gitelson AA, Gritz Y, Merzlyak MN (2003) Relationships between leaf chlorophyll content and spectral reflectance and algorithms for non-destructive chlorophyll assessment in higher plant leaves. J Plant Physiol 160(3):271–282
11. Hennessy A, Clarke K, Lewis M (2020) Hyperspectral classification of plants: a review of waveband selection generalisability. Remote Sens 12(1):113
12. Hu Z, van Belzen J, van der Wal D, Balke T, Wang ZB, Stive M, Bouma TJ (2015) Windows of opportunity for salt marsh vegetation establishment on bare tidal flats: the importance of temporal and spatial variability in hydrodynamic forcing. JGR Biogeosci 120(7):1450–1469
13. Intergovernmental Panel on Climate Change [IPCC] (2014) Climate Change 2014: Synthesis Report; Contribution of Working Groups I, II and III to the Fifth Assessment Report of the Intergovernmental Panel on Climate Change, Core Writing Team. In: Pachauri RK, Meyer LA (eds). IPCC, Geneva, Switzerland, 151 p
14. Intergovernmental Science-Policy Platform for Biodiversity and Ecosystem Services [IPBES] (2019) Global assessment report on biodiversity and ecosystem services of the Intergovernmental Science-Policy Platform on Biodiversity and Ecosystem Services. In: Brondizio ES, Settele J, Díaz S, Ngo HT (eds). IPBES secretariat, Bonn, Germany
15. James D, Collin A, Mury A, Costa S (2020) Very high resolution land use and land cover mapping using Pleiades-1 stereo imagery and machine learning. The International Archives of the Photogrammetry, Remote Sensing and Spatial Information Sciences, Volume XLIII-B2-2020, 2020 XXIV ISPRS Congress
16. Mars JC, Rowan LC (2010) Spectral assessment of new ASTER SWIR surface reflectance data products for spectroscopic mapping of rocks and minerals. Remote Sens Environ 114(9):2011–2025
17. Millennium Ecosystem Assessment (2005) Ecosystems and human well-being: wetlands and water – synthesis. World Resources Institute, Washington
18. Mury A (2021) Évaluation du service écosystémique de protection contre la submersion marine en milieu mégatidal : baie du Mont-Saint-Michel et Côte d'Émeraude. PhD thesis, EPHE-PSL Université, Paris
19. Mury A, Collin A, Houet T, Alvarez-Vanhard E, James D (2020) Using multispectral drone imagery for spatially explicit modeling of wave attenuation through a salt marsh meadow. Drones 4(2):25
20. Mury A, Collin A, Jeanson M, James D, Gloria H, Pastol Y, Etienne S (2020) Mapping nature-based marine flooding risk using VHR wave, airborne LiDAR and satellite imagery: the case study of the Dol March (Bay of Mont-Saint-Michel, France). In: Malvárez G and Navas F (eds), Proceedings from the International Coastal Symposium (ICS) 2020 (Seville, Spain). J Coast Res 95:743–747. Coconut Creek (Florida), ISSN 07490208
21. Norouzi S, Sadeghi M, Liaghat A, Tuller M, Jones SB, Ebrahimian H (2021) Information depth of NIR/SWIR soil reflectance spectroscopy. Remote Sens Environ 256:112315

22. Observatoire national sur les effets du réchauffement climatique [ONERC] (2019) Des solutions fondées sur la nature pour s'adapter au changement climatique – Rapport au premier ministre et au parlement, La documentation française, 306 p
23. Tsakiridis NL, Theocharis JB, Symeonidis AL, Zalidis GC (2021) Improving the predictions of soil properties from VNIR–SWIR spectra in an unlabeled region using semi-supervised and active learning. Geoderma 387:114830
24. Valéry L, Radureau A, Lefeuvre JC (2017) Spread of the native grass Elymus athericus in salt marshes of Mont-Saint-Michel bay as an unusual case of coastal eutrophication. J Coast Conserv 21:421–433

Monitoring Land Cover Change in the Southeastern Baltic Sea Since the 1980s by Remote Sensing

Sébastien Gadal and Thomas Gloaguen

Abstract The political, economic, and social changes associated with the collapse of the Soviet Union at the end of the 1980s led to major land cover and land-use changes in the southeastern Baltic Sea coastal regions. These changes (demilitarization of the coasts, end of collective ownership, specialization of economic activities, etc.) are characterized by a fast process of coastalization with the growth of urban areas, coast suburbanization, and the decrease of agricultural land. At the same time, we observe the implementation of protected natural areas at the regional level and through cross-border cooperation with international organizations (UNESCO, European Union [EU], etc.). Both processes have an important impact on the management of the coastlines of Latvia, Lithuania, and Russia. The analysis of the coastal changes is based on the use of Landsat remote sensing data series from the 1980s to 2020 combined with EU geographic databases and the land-use plans. The comparative analysis of the land cover changes in the Oblast of Kaliningrad, Lithuanian, and Latvian coastal zones allows us to understand the impacts of the three different planning policies since the end of the 1980s. The territorial dynamics are modelled using the GEOBIA package with object-oriented classification and machine-learning algorithms (maximum likelihood, minimum distance to means, parallelepiped classifiers) applied to the Landsat 5 TM and Landsat 8 OLI satellite multispectral images. The produced land cover maps are compared with the Climate Change Initiative Land Cover of the European Space Agency from 1995 to 2015.

Keywords Land cover change · Land planning · Remote sensing · Machine learning · Data accuracy · Southeastern Baltic coast

---

S. Gadal (✉)
CNRS, ESPACE, UMR 7300, Aix-Marseille Université, Université Côte-d'Azur, Avignon Université, Avignon, France

North-Eastern Federal University, Republic of Sakha, Yakutia, Yakutsk, Russian Federation
e-mail: sebastien.gadal@univ-amu.fr

T. Gloaguen
CNRS, ESPACE, UMR 7300, Aix-Marseille Université, Université Côte-d'Azur, Avignon Université, Avignon, France

Kaunas University of Technology, Faculty of Civil Engineering and Architecture, Cultural and Spatial Environment Research Group, Kaunas, Lithuania
e-mail: thomas.gloaguen@ktu.edu

S. Niculescu (ed.), *European Spatial Data for Coastal and Marine Remote Sensing*, https://doi.org/10.1007/978-3-031-16213-8_4

1 Introduction

In recent years, the southeastern Baltic (Fig. 1) has been characterized by significant changes. The land cover maps made using remote sensing show the transformations of the coastal zones of Kaliningrad Oblast, Lithuania, and Latvia. Jūrmala (Latvia) is one of the emblematic examples of a significant illegal real estate investment due to the environmental impact of private Russian citizens destroying the dunes of the coastal zones during the 2010s. Several of the projects built villas directly on the dunes. The number of abandoned houses since the Latvian restoration of independence are the object of strong real estate speculation [2], contributing as well to the gentrification of the coastal zone. Jūrmala is emblematic of this current development in the southeastern Baltic Sea, i.e., a progressive littoralization of their coastal region.

Fig. 1 Definition of the coastal zone studied

The dynamics of coastalization modelled here are common with others observed around coastal lands of the world at the global scale with a high concentration of populations and economic activities [1, 18]. However, the particularity of the southeastern Baltic region is related to the context of political, social, economic, and cultural changes due to the collapse of the Soviet Union [11]. The transition to a market economy led to "a chaotic territorial development" and has made "permanent penetration of intensification of urban process" [19].

Furthermore, the redistribution of the expropriated lands and properties after the Soviet Union induced conflicts between public and private interests. Moreover, due to the inexperience of the inhabitants with land-use regulations, many cases of abuse occurred, especially from the newcomers. They did not respect the legislation, such as limiting access to the coastline, and took part in behaviors that national and international laws try to limit as much as possible, such as reconstructions not adapted to the specificities of the coastal landscape and natural processes, using natural resources, etc., [19]. The spatial analysis modelling of the coastal changes of the southeastern Baltic provides a global diagnostic of the environmental and territorial impacts. It is based on the quantification of land-use changes, using remote sensing Landsat 5 TM and Landsat 8 OLI archives series from 1987 to 2020 based on the GEOBIA approach. The cartographies of the land cover changes are merged with the land-use policies implemented for coastal environmental management in the context of the transition from the Soviet regime to a liberal system.

2 Context

The societal and political concerns about the Baltic coastline vulnerabilities went hand in hand with a growing awareness of the pressures stemming from increasing human activities, which themselves have undergone rapid political, social, and economic change in recent decades [4, 5].

2.1 The Soviet Period: Military Control and Economic Specialization

The coastal territories of the southeastern Baltic Sea of the Soviet Union were heavily controlled; as they were the border territories of the Soviet Union, these territories were militarized, and the restricted access limited the numbers of inhabitants on the shores [18, 19].

At the same time, the Soviet economic system undertook the policy of economic planning specialization with the development of activities related to fishing, tourism, shipbuilding, and the port industry giving the coastal zone a function of an interface [10].

The second investments in terms of economic development concerned agriculture, with the collectivization of lands in the 1940s. In Latvia, the economic and territorial collectivization of agriculture generated massive moves of the populations to the cities, for example [16].

2.2 *Liberalization and Gradual Coastalization of the Southeastern Baltic Countries*

The collapse of the USSR in the early 1990s led to an increasing concentration of economic activities in the coastal regions. The Latvian and Lithuanian coastal zones contribute 5% and 12% of national GDP, respectively [5, 10, 21].

The economic sectors became more specialized. Traditional activities such as small-scale fishing or local farm agriculture units were replaced by heavy industrial manufacturers, ports, and tourism. The opening of the local economy to the Baltic area and the European Union after the countries gained independence accelerated this process [4, 8, 10, 13, 19, 21].

The "economic growth and increase of population are reflected in the landscape by an increase of construction and a decrease of natural territories" [16].

Because of the capacities of Earth observation satellite images to measure the territorial changes, the modelling of land-use change by remote sensing helps to capture these rapid economic, demographic, and territorial transformations.

3 Methodology

3.1 *Definition of the Coastal Zone Study Area*

One of the main problems in coastal research is the delimitation of the coastline and its region of influence. A buffer zone of 100 km was used to delimitate the coastal area from the coastline (marine and lagoon) (Fig. 1).

This method of determining the coastal area was developed by Fedorov et al. [10] in their study analyzing the impacts of the sea interface on the economic and urban development of Kaliningrad Oblast.

The estimated area of sea influence based on their model is 100 km from the coastline. This area of sea influence is calculated spatially from the administrative units used by the European Union (level 3). Some areas are not administratively considered by the EU to be coastal zone territories even though their distance from the coast is only about 50 km.

3.2 Data Acquisition

The spatial land change monitoring is based on the Landsat 5 TM and Landsat 8 OLI satellite multispectral image archives of the Earth Explorer platform (http://earthexplorer.usgs.gov).

Landsat is today the only archive offering free-access, remote sensing data that allows for modelling the land cover change transformations from the mid-1970s till today. It offers a spatial resolution of 30 x 30 m since the launch of Landsat 4 in 1983. The southeastern Baltic coastal zone landscape structures and areas are matched with a spatial resolution of 30 m. The spatial resolution of 30 x 30 m is considered efficient for monitoring all the geographic objects of the coastal zone that organize and structure the landscapes.

The analysis used the 7 bands covering the visible, near infrared, and middle infrared (SWIR 1 and SWIR 2). Because of the spatial resolution of 80 m, we did not select the Landsat 1–3 MSS images from the 1970s to 1983. The Landsat scenes cover the study area from the border of Poland with the Kaliningrad Oblast to the Latvian–Estonian border. To improve the level of land-use object detection and recognition, the selected scenes covered the end of spring, summer, and the beginning of the autumn seasons from May to October. The intensity of light received on the land surface, seasonally, is at its highest during these months of the year. Snow and cloud covers are limited during this period, and the chlorophyll activity improved the capacities of detection among the geographic objects composing the land-use classes.

A temporal interval of 10 years was determined to analyze the land cover change dynamics with a total of 24 scenes (Table 1). Six Landsat scenes are required to cover the entire coastal region studied.

We used Landsat level-1 and level-2 satellite image archives with a calibration level set to surface reflectance (top of canopy), eliminating artefacts related to the atmosphere (e.g., humidity, aerosols), to the intensity of light received on the Earth's surface, and to the geometric offset from the observation area.

3.3 Method of Analysis of Land-Use Change

The land-use changes analysis is based on a GEOBIA approach that aims to study "geographical entities or phenomena through delineating and analysis image-objects rather than individual pixels" [7]. This approach leads us to use a segment-based classification to identify the different modes of land use. This method facilitates the interpretation of land-use changes and allows for a more accurate representation of landscape features than a classification based on the original pixels. Figure 2 summarizes the different steps of the method.

Table 1 Landsat satellite images used in land cover processing

Date of acquisition		Processing date		Mission Landsat	WRS coordinates		Kappa index
Year	Month/day	Year	Month/day		Path	Row	
1987	10/12	2017	02/10	L5	187	020	0.91
						021	0.90
						022	0.94
	10/10				189	020	0.92
						021	0.92
						022	0.93
1997	08/20	2018	02/25	L5	187	020	0.96
						021	0.92
						022	0.93
	05/14		02/19		189	020	0.97
						021	0.97
						022	0.95
2009	08/21	2016	10/22	L5	187	020	0.94
						021	0.92
						022	0.94
	05/31		10/25		189	020	0.95
						021	0.96
						022	0.93
2020	05/31	2020	06/08	L8	187	020	0.91
						021	0.92
						022	0.93
	09/18		10/06		189	020	0.94
						021	0.93
						022	0.94

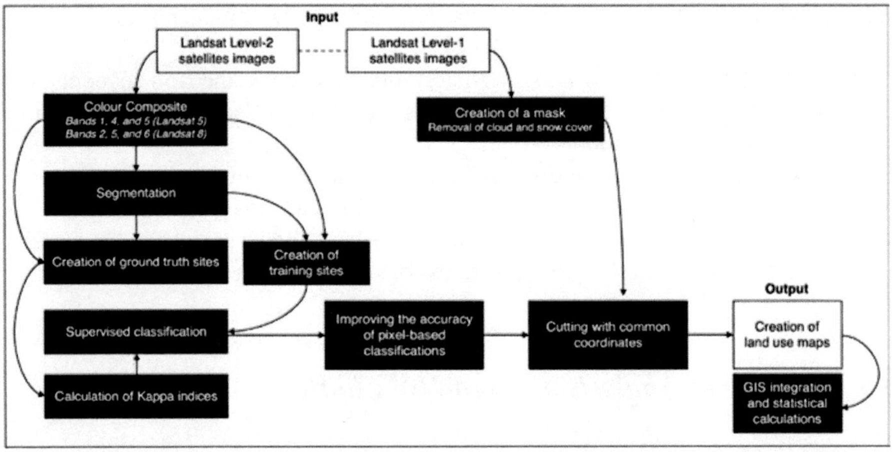

Fig. 2 Methodological scheme of the treatments applied in this study

3.3.1 Pre-processing

The identification of potential cloud cover in satellite image archives is a necessary step to take advantage of the data they offer. For this purpose, we use the F-Mask algorithm from Landsat level-1 images to perform an automated raster containing land surfaces, water surfaces, clouds with the associated shadows, and snow surfaces. These last three elements are extracted and merged into new rasters used as masks for classifications. This operation is applied to each scene with the same WRS coordinates, regardless of the year. This operation allows the calculation of statistics from the classifications based on the same area between the different years of the period analyzed.

3.3.2 Image Segmentation

The image segmentation constitutes the first step of the production of land-use maps/land-use change by object-oriented classification. The segmentation is processed using a color composite combining the near-infrared, mid-infrared 1, and blue bands (Table 2 and Fig. 3). This allows the best identification of the geographical objects of the analysis, i.e., water surfaces, forest surfaces, agricultural surfaces, urban surfaces, and bare surfaces.

These color compositions are used in the segmentation to create object-oriented land uses composed of group pixels sharing a spectral similarity. The algorithm requires a tolerance threshold that defines the level of generalization of the segmentation identifying the land uses. A low similarity tolerance produces homogenous segments, while a high similarity tolerance produces heterogeneous and generalized segmentation (Fig. 4).

3.3.3 Object-Oriented Land-Use Classification

The color compositions were used as a reference for the field survey. The geographic knowledge acquired during the field surveys are contributing to the creation of vector ROI training sites on the segmented images. About 30 ROI training sites were determined for each land-use class, the number of which should vary between 20 and 40 for accurate classification. Then, different supervised classifiers, including parallelepiped, minimum distance, maximum likelihood, Fisher LDA, and k-nearest neighbor (KNN), were tested and compared (Fig. 5).

Table 2 Color compositions used for land-use classification

Canals	Band number	Spectral band	Wavelength
Landsat 8			
Red	5	Near-infrared	0.85–0.88 μm
Green	6	Mid-infrared 1	1.57–1.65 μm
Blue	2	Blue	0.45–0.51 μm
Landsat 5			
Red	4	Near infrared	0.77–0.90 μm
Green	5	Mid-infrared 1	1.55–1.75 μm
Blue	1	Blue	0.45–0.52 μm

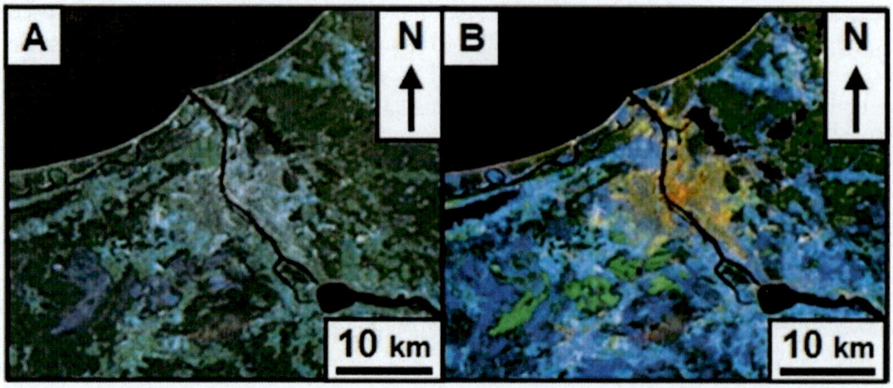

Fig. 3 Color composition of a Landsat image for land cover studies. Example with a Landsat 5 image with "natural color" composition (**a**) and color composition with bands 4, 5, and 1 (**b**). They allow the identification of urbanized surfaces (orange), water surfaces (black), agricultural surfaces (blue and green), forest surfaces (dark green), and bare surfaces (white)

The accuracy of the different classifiers is evaluated by comparing the interpreted image and geographic ROI sites described during the surveys. These geographic sites of accuracy control integrate both field knowledge and GIS databases.

A comparative analysis of the classification methods is performed. The kappa indices of each classifier show that the "maximum likelihood" is the most efficient of them with a value of 0.94 (Table 3). A majority rule algorithm is applied in each segment to determine its class from the classification performed with "maximum likelihood" and to create a segment-based classification.

3.3.4 Post-processing

The common region of interest among the different scenes and dates covering the same study area is extracted based on the same WRS coordinates. The statistics obtained from the classifications are based on the same area and make it possible to compare land use between different years.

4 Results

Based on the segment-based classifications, the statistics of the land-use cover were produced for 1987, 1997, 2009, and 2020. The results obtained show a clear trend of the territorial transformations "footprinted" by the land-use maps despite a variability of values obtained due to the spatial accuracy errors (see Part 5. Discussion). Maps are compared with maps from 1995 to 2015 from the Climate Change Initiative (CCI) Land Cover database of the European Space Agency (ESA) [9] with

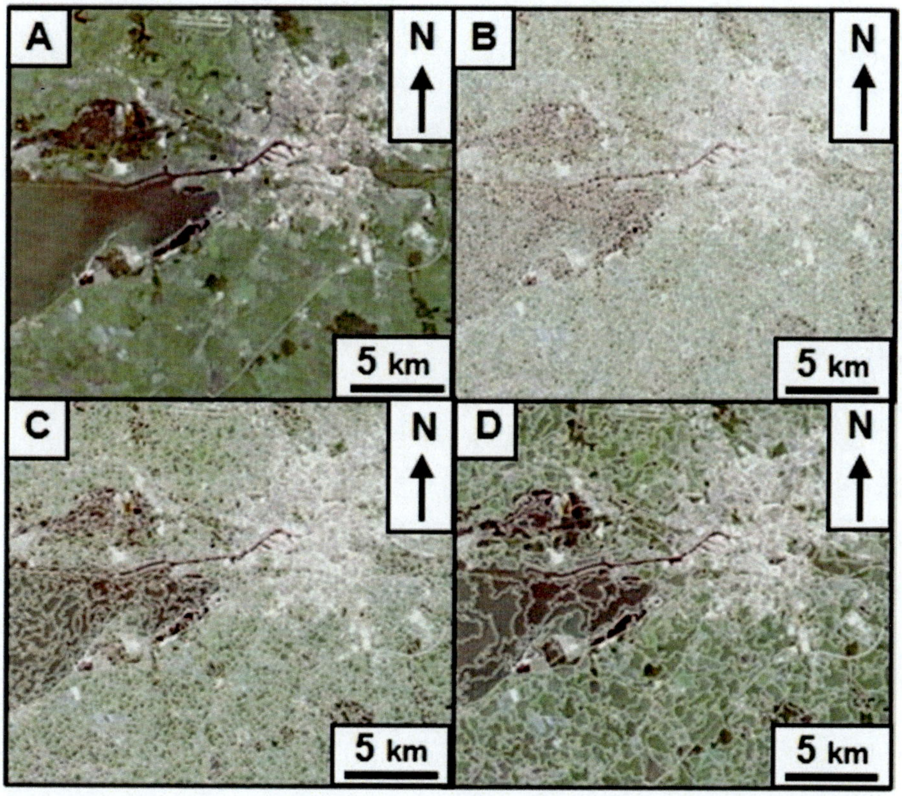

Fig. 4 Examples of different tolerance values used for segmentation. On a "natural color" composition (**a**), segmentation was performed with a tolerance value of 20 (**b**), 50 (**c**), and 100 (**d**). The tolerance value chosen for the study is 20

a resolution of 300 × 300 m. Compared to our land-use classes focused on agricultural, forestry, and urban areas, the CCI Land Cover databases also include other classes such as grassland, sparse vegetation, and wetlands.

4.1 Significant and Distinctive Trends Toward Coastal Urbanization

In Latvia, a clear dominance of urbanized areas can be observed in the zone closest to the shore (Table 4). For example, in 2020, the urban areas represent 13% of the 0–25 km band compared to less than 5% for the 25–50 km, 50–75 km, and 75–100 km bands. In the same way, the CCI Land Cover databases show that in 2015, the rate of urban areas was 1.7% for the 0–25 km band compared to less than 0.5% for the other bands (Table 4). The Latvian coast is "the most densely

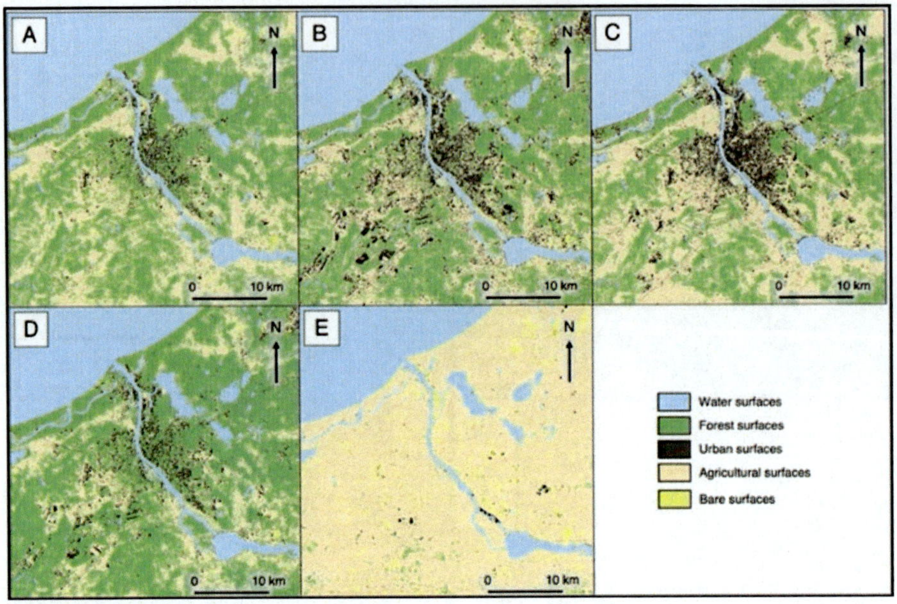

Fig. 5 Extracts of land-use maps obtained using different supervised algorithms. Example of land use in the Riga region (Latvia) with (**a**) Fisher, (**b**) KNN, (**c**) maximum likelihood, (**d**) minimum distance, (**e**) parallelepiped classifications

Table 3 Kappa indices in the supervised classifications used

	1 Fisher LDA	2 KNN	3 Maximum likelihood	4 Minimum distance	5 Parallelepiped
Average of kappa indices	0.91	0.88	0.94	0.88	0.62

populated coastal area in the Baltics" with half of the country's population concentrated there [5, 18]. However, there are still relatively unoccupied areas with low building density, especially in the Gulf of Riga [18].

The results of classification indicate that one-third of the coastal region is occupied by forests (Table 4), although their area has decreased between 8% and 20% for the period 1987–2020. While the trend of decreasing forest area is confirmed by the CCI Land Cover databases (decrease between 6% and 10% for the period 1995–2015), the area of forests could be underestimated as it covers almost two-thirds of the coastal region. Indeed, according to CCI data, the proportion of forest area in each coastal strip varies between 57% and 70% in 1995 and between 51% and 66% in 2015 (Table 4).

On the other hand, agricultural areas have increased slightly: the rates of changes are 3–9% between 1987 and 2020 compared with 10–13% for the CCI Land Cover databases for the period 1995–2015.

Table 4 Share of land-use classes by coastal zoning in Latvia with classification database (1987–2020) and CCI Land Cover databases (1995–2015)

Segment-based classification results (1987–2020)					CCI Corine Land Cover Databases (1995–2015)			
Year	Coastal strip	Forest area	Urbanised areas	Agricultural land	Year	Forest area	Urbanised area	Agricultural land
1987	0–25 km	37.14	11.69	33.78				
	25–50 km	37.63	4.75	46.24				
	50–75 km	32.07	4.16	53.98				
	75–100 km	41.24	4.79	43.07				
1997	0–25 km	29.38	31.14	22.40	1995	69.83	1.12	19.93
	25–50 km	33.24	17.36	34.18		62.04	0.23	30.61
	50–75 km	34.94	7.65	42.99		56.91	0.16	35.95
	75–100 km	46.22	4.12	36.68		69.55	0.48	20.88
2009	0–25 km	33.67	12.12	36.96				
	25–50 km	35.37	2.67	49.34				
	50–75 km	34.68	2.13	52.96				
	75–100 km	47.89	1.85	37.10				
2020	0–25 km	34.10	13.12	34.82	2015	65.63	1.74	22.61
	25–50 km	34.29	4.50	48.01		56.61	0.47	34.59
	50–75 km	28.37	2.90	56.17		51.44	0.30	39.69
	75–100 km	33.17	3.07	47.02		64.07	0.11	23.60

In Lithuania, urbanization is the main geographic trend dynamic observed in the 0–25 km band and the 75–100 km band. Except for the 1987 results, the urban areas vary between 7% and 15% from 1997 to 2020 in the 0–25 km and 75–100 km coastal bands compared to 2% and 6% in the 25–50 km and 50–75 km bands (Table 5). The concentration of urban areas in the coastal region can be explained, in part, by the short length of the coastline—less than a hundred kilometers [21]. However, the increase of urban areas between 1987 and 2020 took place outside the most coastal band (0–25 km). Surface covered has more than doubled for the 25–50 km, 50–75 km, and 75–100 km bands, but it is difficult to have accurate statistics because of the large variability in values observed between these 2 years (Table 5). Moreover, the same trend is visible with the CCI Land Cover databases for the period 1995–2015, which shows that the smallest increase is observed in the 0–25 km band (14% increase between 1995 and 2015 compared to 24–115% for other bands). In general, the coastal area has experienced less urban development than the rest of the country (48% growth for the coastal zone against 88% for the rest of the country between 1995 and 2015).

Forest cover has remained generally stable: the variation is between −1% and 2% for the bands between 0 and 75 km from 1987 to 2009 (2020 shows abnormally low data). The CCI Land Cover databases confirm this relative stability with evolution of 3–9% between 1995 and 2015.

For agricultural land, the classification results show a net decline of between 9% and 23% between 1987 and 2009 (the 2020 data is unusually high). The CCI Corine Land databases show the same trend but with much lower rates of change, 2–4%

Table 5 Share of land-use classes by coastal zoning in Lithuania with classification database (1987–2020) and CCI Land Cover databases (1995–2015)

Segment-based classification results (1987–2020)					CCI Corine Land Cover Databases (1995–2015)			
Year	Coastal strip	Forest area	Urbanised areas	Agricultural land	Year	Forest area	Urbanised area	Agricultural land
1987	0–25 km	69.73	4.13	20.77				
	25–50 km	75.30	1.04	21.36				
	50–75 km	77.04	0.58	19.57				
	75–100 km	19.04	3.04	74.17				
1997	0–25 km	62.21	15.33	17.29	1995	28.46	2.11	59.30
	25–50 km	69.12	6.72	19.77		29.51	0.39	55.91
	50–75 km	70.52	6.84	17.91		35.54	0.44	47.39
	75–100 km	19.14	11.47	62.43		23.23	0.36	67.86
2009	0–25 km	71.14	10.07	15.95				
	25–50 km	74.90	3.59	19.24				
	50–75 km	78.04	2.01	17.23				
	75–100 km	22.18	9.32	64.58				
2020	0–25 km	75.56	7.12	14.68	2015	31.01	2.40	57.08
	25–50 km	30.25	6.14	57.04		30.50	0.49	54.59
	50–75 km	31.66	5.80	53.76		36.62	0.60	45.95
	75–100 km	19.48	10.33	66.10		23.97	0.78	66.67

between 1995 and 2015. Both the classifications and the CCI Corine Land database show that the most significant decline is in the most coastal band.

In the Oblast of Kaliningrad, many urban areas are in the coastal strip. For example, the CCI Land Cover databases show that in 2015, the rate of urban areas was 3% for the 0–25 km band compared to less than 0.6% for the other bands (Table 6). The results of our classifications show the same trend, although the results fluctuate. In 2020, the urban area in the 0–25 km band was more than double any other band (Table 6). Indeed, more than 80% of the population of the Oblast lived in a coastal strip of about 50 km [10].

The most significant changes are realized in the strips near the coastline. Urban areas have almost doubled in the 25–50 km band between 1997 and 2020. The CCI Land Cover databases show that this increase is highest in the 0–25 km band between 1995 and 2015 (Table 6).

Agricultural lands have decreased between 4% and 15% from 1995 to 2015 according to the CCI Land Cover databases, while the decrease is measured between 30% and 65% over the period 1987–2020 according to our classification results.

Forests have increased in area by 23–42% between 1995 and 2015 according to the CCI Land Cover databases, while the variations range from −2% to 53% between 1987 and 2020 according to the classification results.

Clearly, the variability of the data produced from the object-oriented classifications (Table 6) prevents the proposal of robust conclusions. But the results show similar evolutions between the two datasets and are like those observed in Lithuania regarding forest and agricultural areas.

Table 6 Share of land-use classes by coastal zoning in Kaliningrad Oblast with classification database (1987–2020) and CCI Land Cover databases (1995–2015)

Segment-based classification results (1987–2020)					CCI Corine Land Cover Databases (1995–2015)			
Year	Coastal strip	Forest area	Urbanised areas	Agricultural land	Year	Forest area	Urbanised area	Agricultural land
1987	0–25 km	22.80	7.11	58.28				
	25–50 km	70.45	0.91	27.23				
	50–75 km	59.89	2.34	35.63				
	75–100 km	52.76	1.51	44.26				
1997	0–25 km	17.58	24.91	44.95	1995	29.98	1.67	56.97
	25–50 km	66.05	5.20	25.88		19.29	0.35	74.66
	50–75 km	59.61	6.89	30.81		16.33	0.34	78.70
	75–100 km	64.63	2.02	31.91		19.01	0.12	76.70
2009	0–25 km	27.00	19.47	40.74				
	25–50 km	78.70	2.43	16.84				
	50–75 km	73.55	3.46	21.49				
	75–100 km	68.96	0.74	29.02				
2020	0–25 km	23.46	27.03	41.04	2015	36.95	3.32	48.56
	25–50 km	69.34	10.56	18.81		27.30	0.67	66.33
	50–75 km	69.17	10.57	18.91		21.20	0.47	73.90
	75–100 km	80.79	2.24	15.52		21.88	0.17	73.53

4.2 *Evolution of Land Use, Territorial Development, and Planning Policies*

Results from land-use classifications and CCI Corine Land Cover databases provide clear trends that highlight the importance of planning and development policies and their effectiveness.

The legislation during the Soviet Union, created in the early 1980s, aimed to control the "location of recreational places for the national economy objects and arrangement of such territories […]", thus allowing a balance between environmental protection and the development of economic activities [16].

Since the restoration of independence, the legislative framework for environmental protection, especially in coastal areas, has evolved differently in the Baltic countries and the Oblast of Kaliningrad.

In Lithuania and Russia, we observed a stabilization or increase of the forest cover, while in Latvia, there has been a decrease in recent years (see Sect. 3.1). In the latter, this loss could be associated with "excessive exploitation of natural resources" caused by agricultural and forestry activities, despite considerable progress in environmental regulation, particularly since Latvia's accession to the European Union [14, 17]. The naturally protected coastal areas are also the main tool for environmental preservation, but the lack of management plans and insufficient funding limits the effectiveness; forestry contributes to more than 6% of Latvia's GDP but only half of the country's forests are certified for sustainable management, which is lower than the averages of other OECD countries with a high forest cover [17].

In Lithuania, natural processes such as reforestation of wetlands, mainly swamps, in part, explain these results [15]. However, this growth of forest cover is mainly related to the forest "recovery" programs: forestry companies engage in replanting when they clear land, which led to "forest management in Lithuania [...] rather intensive and sustainable during the last decades" [15, 20, 21]. The forest growth has been at the expense of abandoned agricultural plots [15].

This decline in agricultural land has been observed since the 1950s following policies of nationalization and collectivization of the land [15]. However, since the restoration of independence, the agricultural sector has lost a lot of its economic value: This has resulted in a decrease in the number of agriculture workers, and the abandonment of farms and land fields [19]. In Lithuania and in the Kaliningrad Oblast, these dynamics are observed through remote sensing and the CCI Land Cover databases [15, 21].

The decrease in agriculture led to the development of second homes, which has resulted in the increase of urban areas in the coastal strip for the three countries [19]. Most of the urban areas in Latvia and Russia are close to the coastline. Considering only the coastal zone, the classification data indicates that 48% and 63% of urban areas are concentrated in the 0–25 km band in Russia and Latvia in 2020. In comparison, the CCI Land Cover databases show that this rate reaches 80% and 70% in 2015. This concentration is due to the presence of large cities like Riga, Jūrmala, Kaliningrad, and Baltiysk. While an increase in urban areas has been observed, spatial variations of this extension can be also observed.

The weak legislation due to "lack of definition of the main pressures on biodiversity" [17] could explain that urbanization is spreading in the most coastal strip of Latvia and Russia, reflecting an "uncontrolled construction of buildings on the seashore, in connection with the development of tourism activities" [14].

Compared to the Oblast of Kaliningrad (Russia) where there is no independent legislation specific to coastal areas (the source of multiple conflicts of interest), the increase of urbanized areas in the most coastal strip in Latvia is less significant. It reflects the relative effectiveness of protected areas, present on half of the coastline, as a brake for urbanization growth [1, 3, 16].

However, the development of protective legislation has been eventful and has allowed "the illegal use of natural resources and quite often also degrading of natural territories." Moreover, the entry into the European Union (2004) allowed the implementation of new planning legislation with the follow-up of European environmental directives [5, 16].

In contrast, the increase of urbanized areas is mainly located outside the 0–25 km band in Lithuania (see Sect. 3.1) with a regular and continuous growth, showing a "well-developed legal basis for the spatial planning" in this country [20, 21]. In general, the Lithuanian legislation is well developed in giving priority to the protection of natural coastal processes and landscapes [13, 19, 22].

Several laws define the management and use of coastal zones of Lithuania. The Law of Coastal Strip (2002) is based on the principles of sustainable development:

the development of economic activities while ensuring the protection of the environment and exploited natural resources [22]. The aim is to limit construction within a hundred-meter strip from the coastline and to regulate work on properties, their renovation, and restoration; the coasts are state property whose anthropic activities must be regulated [19].

Another important law is the Law on Spatial Planning (1995). It aims to protect emblematic landscapes and biodiversity, as well as cultural and natural heritage by promoting "sustainable development" and "reasoned urbanization" of the Lithuanian territory. This law defines the compatibility of the various urban planning documents and the "requirements of the decision of territorial planning process" [12].

In the southeastern Baltic countries, the implementation of specific legislation for coastal zones has been influenced by the development of international strategies and programs such as the Helsinki Convention (definition of environmental protection objectives) or the recommendation of the European Parliament and the European Council concerning the implementation of Integrated Coastal Zone Management, which implies an articulation of the different planning tools and documents between multiple actors and scale of intervention [13].

5 Discussion

5.1 Limitations of Coastal Zone Management Policies

In Lithuania, the environmental protection laws are considered too restrictive for the economic activities in protected areas [19]. Planning policies based on "integrated" management are difficult to apply when public and private interests are divergent, especially in unprotected areas of the coastline such as in Giruliai or Melnragė (Lithuania), where the interests of private landowners override existing national and local legislation [19].

These policies require significant financial, technical, and human resources to be fully effective, but these are currently still insufficient [22].

For example, former military sites and collective farms were privatized in the early 1990s in Latvia with independence. In the absence of management strategies, there have been many cases of abuse. Insufficient funding did not allow the state to acquire these lands, which were degraded despite providing minimal environmental protection during the Soviet period [5, 16]. In Lithuania, expenditure to preserve coastal dunes amounts to almost 15,000 euro per kilometer of coastline and justifies limited action on critical sections only [19].

Similarly, in Russia, the lack of funding leads to an increasing number of human settlements that are now vulnerable to erosion, for example [6].

5.2 Limits of Supervised Classification by GEOBIA Approach

The main analysis of the dynamics of the coastal changes was made using remote sensing and the CCI Land Cover databases.

Numerous spatial accuracy errors were identified due to the use of automated processing to carry out the land-use classification.

The spatial accuracy errors identified are at the origin of significant variations in the results of land-use cartographies and statistics generated despite kappa indices greater than 0.90. These data do not allow us to determine the fine impact of the end of the Soviet system on the evolution of land use, especially in the first years after independence.

These errors of accuracy to the land field reality are linked to difficulties in identifying certain land-use classes, particularly the smallest urban entities (villages) and suburban residential areas because of the Landsat 5 TM and Landsat 8 OLI spatial resolution of 30 × 30 m (Fig. 6).

The spatial resolution of the Landsat 5 TM satellite images limits the level of detection of isolated habitats not covering an area of 900 m^2. The use of the panchromatic band with a resolution of 15 m of Landsat 8 OLI images can improve the spatial accuracy level of the urban areas. However, to standardize the processing and the comparability of the land use produced among the dates, pansharpening (fusion of multispectral data of 30 m with the panchromatic band of 15 m spatial resolution), possible only with Landsat 8 OLI data, was not applied.

Despite the use of remote sensing data in the chlorophyll optimum (spring and summer) for the improvement of the spectral separability among geographic objects such as the urban areas and agricultural lands, spectral confusions persist. They are visible between bare soil and some agricultural plots (Fig. 7), between some agricultural plots and urban areas (Fig. 8), and between some forest and agricultural plots.

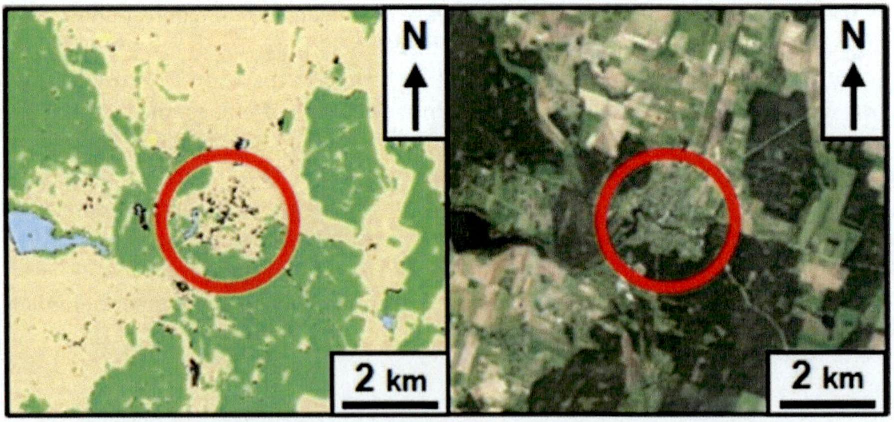

Fig. 6 Difficulties in identifying the village of Darbėnai (Lithuania)

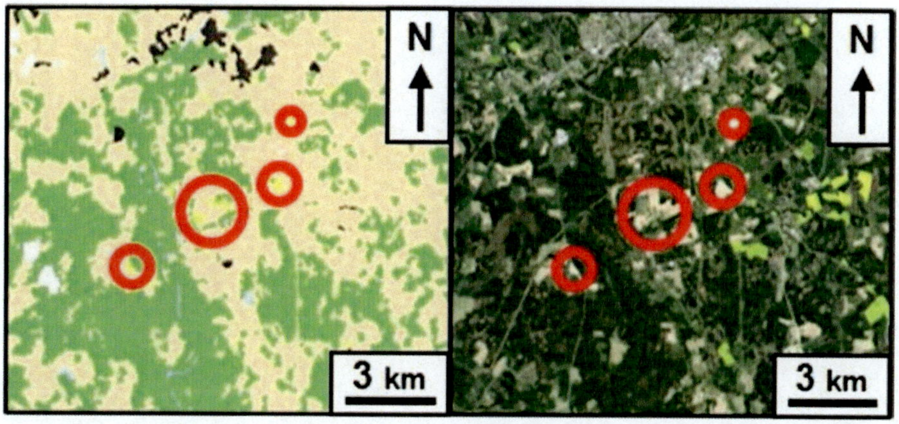

Fig. 7 Spectral confusions with agricultural plots considered as bare soil

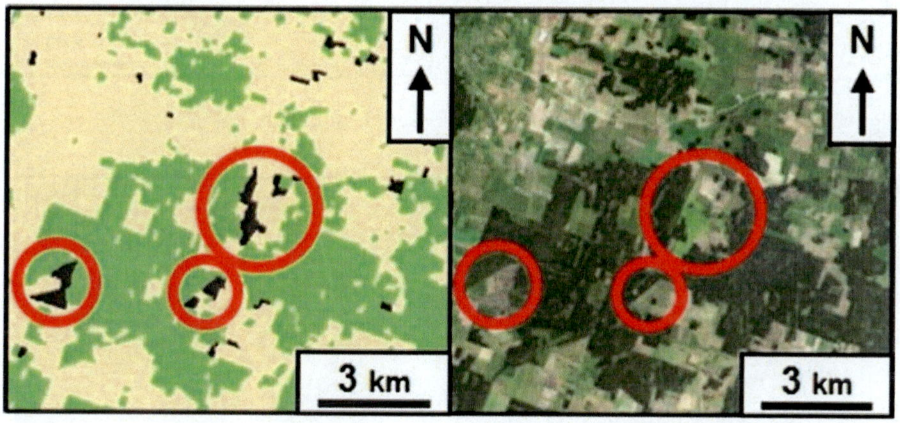

Fig. 8 Spectral confusions with agricultural plots considered as urbanized areas

The use of the F-Mask algorithm was not sufficient to eliminate the cloud cover from the satellite images. The clouds and their shadows are the cause of multiple spectral confusions with urbanized areas, which are overestimated on certain dates of satellite image acquisitions (Fig. 9).

The choice of segment-based classification with the use of a majority rule algorithm to smooth the results in segments causes a significant loss of information for certain land-use classes such as urban areas, for example, especially in peri-urban areas (Fig. 10).

Furthermore, despite the use of Landsat level-2 satellite images, the differences linked to the radiometry and the atmospheric properties of the satellite images did not allow for a homogeneous radiometric result (Fig. 11), making it difficult to discriminate and detect each type of geographical object.

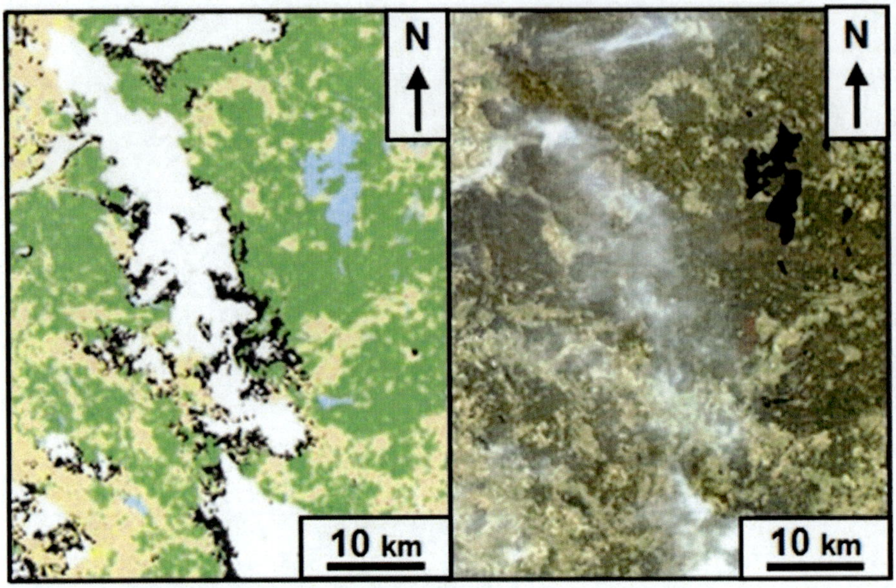

Fig. 9 Incomplete removal of cloud cover in supervised classification

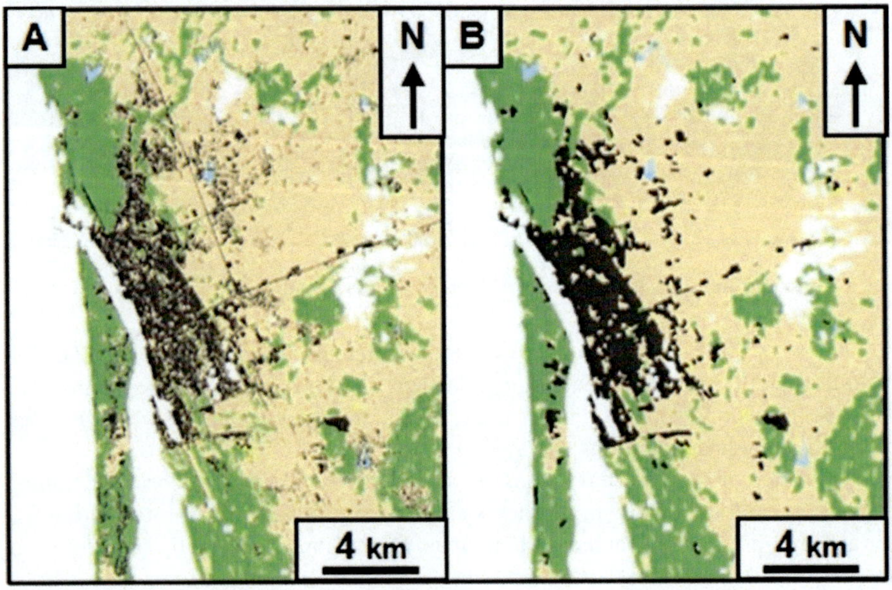

Fig. 10 Segmentation effects on supervised classifications

Fig. 11 Differences between atmospheric and radiometric properties of two separate Landsat scenes

6 Conclusion

The study of the land cover changes between 1987 and 2020 in the coastal regions of the countries of the southeastern Baltic Sea has highlighted a generalized phenomenon of coastalization, characterized by the progressive concentration of population and economic activities on the coastal zone, especially since the end of the Soviet Union.

We performed classifications based on the maximum likelihood classifier after comparing the performance of different supervised classification methods for the years 1987, 1997, 2009, and 2020 from Landsat 5 TM and Landsat 8 OLI satellite image archives. The use of the majority rule algorithm produces segment-based classifications for a more accurate representation of landscape features. The results were compared with the CCI Land Cover databases from 1995 to 2015.

They showed a stabilization in agricultural areas in Latvia but a decline in Lithuania and in the Kaliningrad Oblast, which is associated with the privatization of land and the decreasing economic interest in agriculture.

Economic activities related to tourism and the naval industry have benefited from the intensification of international trade caused by the transition to a liberal system. This movement has gone hand in hand with significant development of urbanized areas characterized by peri-urbanization through the appropriation of second homes, workers' housing, or the development of illegal buildings.

The natural areas such as forest areas grew, and they are dependent on the implementation of management and protection policies to safeguard them. The protection policies and implementations may differ among each Baltic state and the Oblast of Kaliningrad (Russia) due to the legislative, technical, and legal "vacuum" after the end of the Soviet Union. This "vacuum" has been resolved, thanks to the accession to the European Union, but also by the following of several international conventions and directives promoting sustainable development, reconciling environmental protection and economic activities: the Helsinki Convention, recommendation of the European Parliament and of the Council on Integrated Coastal Zone Management, etc.

Finally, this study has highlighted the usual limits of automated remote sensing techniques, which reveal problems in the recognition and detection of certain geographical objects, spectral confusions, and errors related to the processing of images with different radiometric and atmospheric properties, etc. Segment-based classifications reflecting a GEOBIA approach show limits in terms of modelling the smallest urban entities (peri-urban housing, secondary houses, allotments, etc.). This method will undoubtedly demonstrate its relevance with the use of satellite images with a resolution of about ten meters, such as the Sentinel 2 MSI images.

Acknowledgments This research is supported by the CNES AICMEE TOSCA program (Apport de l'Imagerie satellitaire Multi-Capteurs pour répondre aux Enjeux Environnementaux et sociétaux des socio-systèmes urbains).

References

1. Afanasyeva EP (2014) Characteristics of tourism development in the coastal areas of the Kaliningrad region. J Reg Stud 18
2. Aubin L (2013) Le tourisme russe à Jūrmala: l'appropriation territoriale de la périphérie balnéaire de Riga. Regards sur l'Est. http://regard-est.com/le-tourisme-russe-a-jurmala-lappropriation-territoriale-de-la-peripherie-balneaire-de-riga. Accessed 15 June 2021
3. Bass O (2015) Modern conception of coastal protection and problems of hydraulic engineering in Kaliningrad region. Vestnik IKBFU (1)
4. Bitinas A, Žaromskis R, Gulbinskas S, Damustyė A, Žilinskas G, Jarmalavičius D (2005) The results of integrated investigations of the Lithuanian coast of the Baltic Sea: geology, geomorphology, dynamics, and human impact. Geol Q 49(4):355–362
5. Brunina L, Rivza P, Konstantinova E (2011) Coastal spatial planning problems in Latvia. J Coast Res:1224–1227
6. Burnashov E (2011) Current coastal dynamics in the Kaliningrad region, base on annual monitoring surveys. Issues of modern science and practice. Vernadsky University. IV Vernadsky (2), p 10–17
7. Chen G, Weng Q, Hay G, He Y (2018) Geographic object-based image analysis (GEOBIA): emerging trends and future opportunities. GIsci Remote Sens:159–182
8. Eaglet V (1999) Environmental problems of the Baltic Sea and the Kaliningrad region
9. European Space Agency Climate Change Initiative, Land Cover project (2017) 300 m annual global land cover time series from 1992 to 2015. Retrieved from https://www.esa-landcover-cci.org/?q=node/175
10. Fedorov G, Kuznetsova T, Razumovskii VM (2017) How the proximity of the sea affects the development of economy and the settlement in Kaliningrad Oblast. Reg Res Russ 7(4):352–362
11. Gadal S (2010) Urban dynamics in Kaliningrad and Klaipėda coastal regions by remote sensing and GIS. Bridges/Tiltai 50(1)
12. Gaudėšius R (2015) Influence of general plans on urbanization of agrarian territories on Lithuania's seaside. Environ Res Eng Manag 71(4):19–27
13. Gulbinskas S, Suzdalev S, Mileriene R (2009) Coastal management measures in Lithuania Baltic coast (South-eastern Baltic). Coast Eng 5:4042–4052

14. Kraus H, Meyer T (1998) Environmental policy in Latvia. European Parliament
15. Lekavičiūtė J, Gadal S (2009) Forest cover changes during 1975–2005 in the Lithuanian Coastal Region. Rural Dev 2008:241
16. Nitavska N, Zigmunde D (2013) The impact of legislative rules and economic development on the coastal landscape in Latvia. Civil Eng 13:259
17. OECD (2019) OECD environmental performance reviews: Latvia 2019 (short version). OECD Environmental Performance Reviews, OECD Publishing
18. Pranzini E, Williams A (2013) Coastal erosion and protection in Europe. Routledge
19. Spiriajevas E (2014) Hindrances and suggestion for sustainable development of Lithuanian Coastal Strip (Zone). Reg Form Dev Stud 6(1):125–136
20. Vaitkus G, Vaitkuvienė D (2005) Land cover changes in the Lithuanian coastal zone during 1975–2000. Acta Zool Litu 15(2):183–187
21. Veteikis D, Šabanovas S, Jankauskaitė M (2011) Landscape structure changes on the coastal plain of Lithuania during 1998–2009. Baltica 24(2):107–116
22. Žilinskas G (2008) Distinguishing priority sectors for the Lithuanian Baltic Sea coastal management. Baltica 21(1–2):85–94

Assessment of Land Cover Changes in the Allala Watershed Based on Object Based Image Analysis Using Landsat and Sentinel-2 Images

Narimane Zaabar, Simona Niculescu, and Mustapha Kamel Mihoubi

Abstract The coastal city of Ténès, located in northwestern Algeria, is exposed to several natural hazards, such as floods, earthquakes, landslides, and forest fires. Due to human activities, socio-economic constructions, agricultural activities, and the resulting population acceleration, land cover and land use (LULC) dynamics in the city are changing over time. Hence, the understanding of LULC changes and its interactions with human activities and natural hazards is essential for appropriate land management and decision-making. In this study, we investigate LULC changes in the Allala watershed, including the city of Ténès, using remote sensing methods and Geographic Information System (GIS) tools. Object-based image analysis (OBIA) based on random forest (RF) and support vector machine (SVM) machine learning algorithms was performed to provide LULC classification maps, and then, LULC changes were assessed using GIS. In order to assess LULC changes, we used three images acquired using remote sensing, corresponding to 3 years; 1999, 2009, and 2020. A Sentinel-2 image and two Landsat images were used as input data in our methodology. Our LULC classification results showed that RF outperformed SVM on the three input data periods, with an overall accuracy of 95.6% obtained with the Sentinel-2 image. Given the changes over time, it is clear that the Allala watershed has undergone significant changes over the years, particularly an increase

---

N. Zaabar (✉)
University of Western Brittany, CNRS, LETG Brest UMR 6554 CNRS, Technopôle Brest-Iroise, Plouzané-Brest, France

ENSH, National Higher School of Hydraulics, MVRE, Laboratoire de Mobilisation et valorisation des ressources en eau, Blida, Algeria

S. Niculescu
Laboratory LETG-Brest, UMR 6554 CNRS, University of Western Brittany, Plouzané, France
e-mail: simona.niculescu@univ-brest.fr

M. K. Mihoubi
ENSH, National Higher School of Hydraulics, MVRE, Laboratoire de Mobilisation et valorisation des ressources en eau, Blida, Algeria
e-mail: n.zaabar@ensh.dz

S. Niculescu (ed.), *European Spatial Data for Coastal and Marine Remote Sensing*, https://doi.org/10.1007/978-3-031-16213-8_5

in building infrastructure and agricultural land due to population and urbanization growth. Analyzing and mapping the trends of LULC changes in the study area provide a basis for strategic planning and managing, and results of LULC changes can be used as a decision support tool and provide further help in regional and national land management.

Keywords LULC changes · Allala watershed · Landsat images · Sentinel-2 · Object-based image analysis (OBIA) · RF · SVM

1 Introduction

Land use and land cover (LULC) is considered a critical environmental issue with global implications in environmental management and sustainable development [16, 32]. In fact, due to natural and artificial factors, LULC around the world is still undergoing considerable changes, most notably changes due to climate change, rapid urbanization (mainly in critical areas), population growth that requires urban agglomeration development, and the subsequent construction land expansion.

The coastal town of Ténès, located in northwestern Algeria, has observed an increase in changes in LULC due to diverse causes. Indeed, the region is exposed to ongoing natural disasters, such as earthquakes and landslides [2, 29], and natural hazards such as landscape degradation and flash floods [22] due to climate changes, which highly modify LULC dynamics because of the hydrological process of flash floods [13, 14]. At the same time, the region experiences significant deforestation and many forest fires, such as the one in 2014 [7]. As a part of a sustainable development strategy and because of its privileged location, both historical and touristic, the city has integrated several projects based on human activities, socio-economic constructions, and agricultural activities. This implicates population growth and urban construction, as well as the evolution of agricultural areas and, in consequence, population acceleration, as factors that directly lead to changes in the LULC trends of the city over time. In this context, rapid data acquisition and detection of LULC changes are an essential element in environmental monitoring, urban planning, and sustainable development. Additionally, the understanding of LULC dynamics and changes, and its interactions with human activities and natural hazards, is essential for appropriate land management and decision-making improvements [37]. Furthermore, LULC changes and evolution information is considered critical for several environmental considerations, such as water resource management and natural hazard assessment [6, 27, 36, 40].

In recent decades, remote sensing has become more widespread in the scientific field and has emerged as a useful way to track LULC changes based on LULC classification techniques [10, 18, 24]. Pixel-based image analysis methods have been the most widely used to produce LULC maps [25, 30]. These methods consider only the spectral characteristics of the input image, which implies certain limitations of the produced classifications of LULC and any trend detection. These limitations may be partially overcome by considering other image features in order to detect LULC classes with high accuracy [8, 9].

Consequently, the use of object-based image analysis (OBIA) and machine learning classifiers has emerged in the remote sensing community as a way to better address LULC classification and change detection. This method considers the spectral, textual, and contextual information of pixels. Recently, OBIA classification has now replaced conventional pixel-based methods and will facilitate land cover classification using high spatial resolution remote sensing imagery [8, 17, 38]. In addition, according to previous studies, this method has successfully provided the accurate classification of LULC changes using high- and very-high-resolution images [1]. provided LULC maps using Landsat images for 1985, 1990, 2000, 2007, and 2014 for five cities of Saudi Arabia. The aim of their study was to assess the urban growth in these cities. LULC classification was carried out using an OBIA approach. The classified images were also used to predict LULC changes and the growth of urban areas for 2024 and 2034 using specific models. Changes were assessed through transition probabilities. Extreme Gradient Boosting (XGBoost)-based informative feature selection and the random forest algorithm were used to ensure the OBIA classification. The results of LULC classification showed higher values of the overall accuracy (OA), up to 90%, which was very beneficial in assessing LULC [20]. applied the OBIA method in order to classify LULC and evaluate the changes in the Cameron Highlands in Malaysia by taking advantage of the OBIA considerations of texture, shape, position, and digital number, as well as a series of band combinations. As remote sensing data, the authors used Landsat time series images from 2009 to 2019. Similarly, [28] mapped LULC changes in the core zone sand dune located in Indonesia using aerial images taken between 2015 and 2020. The nearest neighbor algorithm was used for LULC classification after a segmentation process. Analysis of the land use changes was carried out by comparing the land use classification results of 2015 and 2020.

The aim of this study is the assessment of the spatio-temporal patterns of LULC across the Allala watershed over a 21-year period (1999–2020), based on remote sensing methods and GIS applications. First, OBIA classification–based machine learning algorithms (random forest and support vector machine) were employed to map LULC over 3 years: 1999, 2009, and 2020. Then, based on the more accurate LULC maps, the assessment of LULC changes was conducted using post classification analysis implemented using GIS applications.

2 Study Area

The study area is the Oued Allala watershed located in northern Algeria on the Mediterranean coast. The watershed covers a total area of 307 km^2 with a length of 35 km for its principal thalweg (Fig. 1). The area is situated between the maximum and minimum altitude, 989 m and 0 m, respectively, and includes Ténès City, a tourist and port city and the second largest city in the Chlef Wilaya. The region is

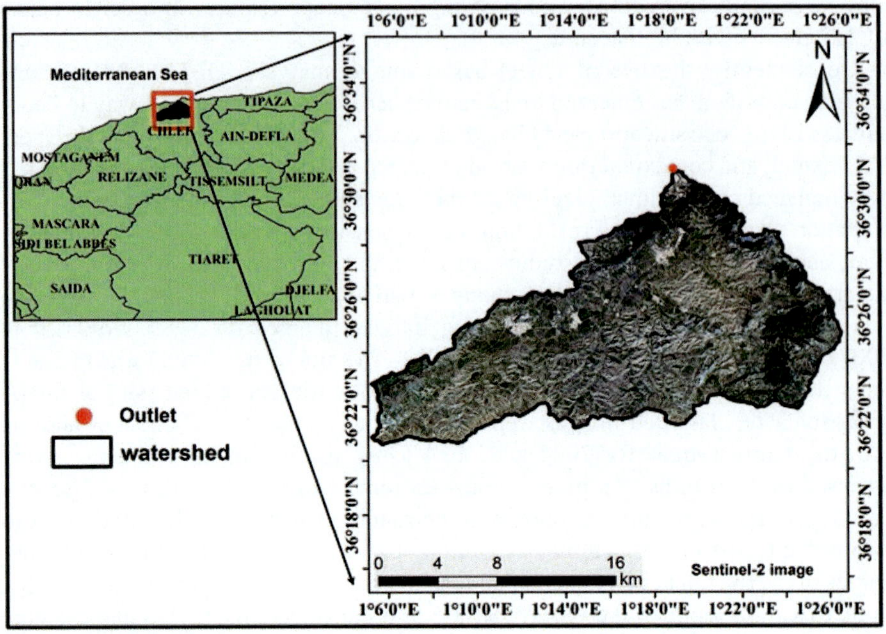

Fig. 1 Location of the study area

influenced by the Mediterranean climate and is characterized by a warm and temperate climate with significant precipitation in the winter months [39]. The average temperature is 18.6 °C and the average total annual rainfall is around 585 mm [22]. The Oued Allala watershed is an area with a high degree of landscape diversity, including the presence of human settlements, forests, and agricultural areas. The northern parts of the Allala catchment are covered by pine forest. Annual crops such as cereals cover the southern part of the watershed. This diversity in LULC classes necessitates an accurate classification and the monitoring of LULC changes.

In addition, as a part of the national strategy to combat natural hazards, specifically floods, the Ténès region was chosen for this study because of its regional importance, its geographical location, and its elevated risk of natural hazards, mainly floods. The analysis of LULC in this area is critical input information for flood vulnerability assessment.

3 Material and Methods

Through a consecutive process, we mapped and assessed land cover/use changes in the Allala watershed for over 21 years. Remote sensing methods combined with GIS applications were both used. Both Landsat and Sentinel-2 data were acquired to establish the classification following two essential steps: (1) The object-based

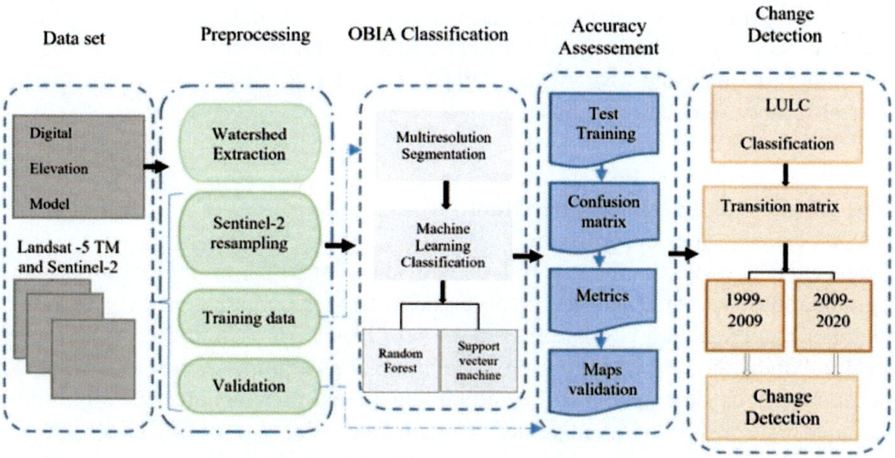

Fig. 2 Workflow of the proposed methodology

image analysis (OBIA) method was applied to provide land cover/use maps with two machine learning algorithms, random forest (RF) and support vector machine (SVM). (2) Then, the assessment of LULC changes was conducted in ArcMap software (version 10.8.1) by post-classification analysis and transition matrix generation (Fig. 2).

3.1 *Object-Based Image Analysis (OBIA)*

3.1.1 Data Acquisition and Preprocessing

Two different types of remote sensing images were acquired from the Landsat and Sentinel-2 sensors to detect LULC changes and facilitate classification. The two Landsat-5 Thematic Mapper (TM) images, with 30 m of resolution, were acquired on April 10, 1999, and May 8, 2009. Landsat-5 TM data are available with free access from the USGS Global Visualization Viewer (https://glovis.usgs.gov). The Landsat-5 TM images were already corrected.

The third image used is a Sentinel-2 image acquired on March 8, 2020, with atmospheric correction. Sentinel-2A contains 13 spectral bands, including four bands with 10 m spatial resolution, six bands with 20 m spatial resolution, and three bands with 60 m spatial resolution. A resampling into a spatial resolution of 10 m was applied in the Sentinel platform (SNAP) software before using the image in the classification process [37]. The dates were selected based on image quality (absence of cloud cover), availability, and the same season for the 3 years (1999, 2009, and 2020), in order to better detect LULC changes. Auxiliary data, in the form of a digital elevation model (DEM) with 30 m resolution, was acquired to highlight the Oued Allala watershed.

Based on very-high-resolution images, training and validation data were generated to process the classification. According to the Landsat-5 TM image resolution and study area analysis, five LULC classes were defined: built-up, forest, roads, cultivable lands, and barren lands.

3.1.2 Multi-resolution Segmentation

Because OBIA considers the spectral, textual, and contextual information of pixels, segmentation is an important step in this process. The segmentation process regroups neighboring pixels with similar characteristics. Indeed, for each image, using input spectral bands, the multi-resolution segmentation algorithm [4] was used to generate homogenous objects. Multi-resolution segmentation is a powerful algorithm designed to iteratively segment a satellite image into objects according to conditions imposed by the user [5]. In our case, using Trimble's eCognition Developer 10.0, we applied a multi-resolution algorithm to the three input images to provide segmentation layers. The identification of objects using this algorithm is built upon relative image object homogeneity or heterogeneity, based on spectral and shape criteria [12]. The size of objects is set by a scale parameter (in this study, the scale parameter = 30 for Landsat images and 10 for the Sentinel-2 image). Heterogeneity of objects is defined by shape and color parameters, in which their proportion is specified by the shape parameter (in this study, shape parameter = 0.1), which means that heterogeneity is influenced 10% by shape and 90% by color) [23, 26]. Similarly, the shape parameter is defined by two components, compactness and smoothness, in which their proportion is specified by the compactness parameter (in this study, compactness parameter = 0.9), which means that heterogeneity is influenced 90% by compactness and 10% by smoothness.

3.1.3 OBIA Classification–Based Machine Learning Classifiers

Machine learning algorithmes are widely used in remote sensing community to LULC change detection [11, 17, 33] based on multiple algorithms, namely, RF, SVM, and maximum likelihood, are widely used to detect changes in LULC. In our study, machine learning algorithms were taught the classification step through generated training samples data. Two classifiers were applied in this stage, RF and SVM. RF is a non-parametric algorithm that performs on multiple decision trees. Each decision tree is constructed using a bootstrap sample driven by different subsets. Each unique set of trees is then applied in order to classify the image, resulting in the final classification, which is a collection of multiple trees and which assigns classes by majority voting. The RF algorithm is simple to execute and only two parameters need to be set up: the number of trees and the number of features in each split. Regarding classification using the SVM algorithm, it is based on the linear function kernel. The principle of this algorithm is to find a hyperplane that separates two classes. The values closed to the hyperplane are the support vectors. The two essential parameters of this algorithm are parameter C that controls the complexity of the classifier and parameter γ that controls the number of carrier vectors to obtain the best hyperplane [21].

3.2 LULC Map Validation

The validation of LULC maps is an indispensable step that confirms the precision of the used methods and the possibility of their use for LULC change assessments. The validation of LULC classified images was conducted through the confusion matrix using validation data. From the confusion matrix, we calculated the coefficients typically utilized in accuracy assessments, the kappa index, and the overall accuracy (OA). The producer's accuracy (PA) and user's accuracy (UA) of the LULC classes were also derived. The PA for a given LULC class shows the probability that a pixel assigned to that class in the ground data will be assigned to that class in the LULC map. The UA shows the conditional probability that a pixel classified into this class in the LULC map will be classified into this class in the ground data [35].

3.3 LULC Change Detection

Based on the three final LULC maps of 1999, 2009, and 2020, changes were assessed using the MC method implemented in Arc GIS used to generate the transition matrix of LULC classes. The MC model is a stochastic process [31, 34] that assigns the probability of the transition of land cover classes from one class to another. LULC changes for the studied area were recorded over two periods, 1999–2009 and 2009–2020. At the same time, a transition matrix was calculated for both periods using a combination of two classified images. A transition area matrix was also computed using a probability matrix. Area transition represents the total area (in cells) forecasted to change from one LULC class to another over the prescribed number of time units [33].

4 Results

4.1 Accuracy Assessment of LULC Classification

The accuracy assessment results of LULC classifications indicate high overall accuracies for both the Sentinel-2 and Landsat-5 input images. Table 1 illustrates the results of the overall accuracies (OA) and the kappa coefficient of the LULC classified images for 1999, 2009, and 2020. As well, when the algorithms were compared, higher accuracies were obtained with the RF algorithm applied to the Sentinel-2 image (2020), with an OA of 96.6% and kappa coefficient of 0.95. In addition, LULC classification with the Landsat image (1999) also resulted in a high accuracy, with an OA of 96.2% and a kappa coefficient of 0.95. Regarding the SVM algorithm, the OA and kappa results were also relatively high, with the higher result obtained using the 2009 Landsat image, with an OA of 94.9% and kappa of 0.93.

Regarding LULC class accuracies (UA and PA) (Table 1), with respect to the 2020 LULC classification–based RF classifier, the UA and the PA of cultivable land, roads, and barren land were both high at more than 98%. However, built-up and forest were relatively poorly classified in terms of UA and PA accuracies. Moreover, for the 2009 LULC classification, built-up and forest were well classified in terms of UA (97.8% and 100%, respectively). Cultivable lands and roads had UA values of 94.6% and 92.7%, respectively. Similarly, for the 1999 LULC classification, forest, roads, and barren lands were highly classified in terms of UA, with 99.8% for both forest and roads and 99.9% for barren land.

Considering SVM classifications over the 3 years, UA values were significantly less than RF values. As well, for the 2020 LULC classification, cultivable land and roads were highly classified in terms of UA, with 90.7% and 86.8%, respectively. However, built-up and barren land were relatively poorly classified; the UA was 54% for built-up and 77% for barren land. For 1999, the LULC classification using the SVM algorithm, forest and barren lands were highly classified, with UA values of 90% for forest and 85.6% for barren land. For the 2009 LULC classification, forest and cultivable lands were well classified compared to other classes, with an UA up to 99% for both classes. Cultivable land and roads for this year stand as poorly classified, with a low UA value (up to 66%).

Overall, comparing both remote sensing data types used to produce LULC maps based on machine learning classifiers (RF and SVM), RF gives higher results, in

Table 1 Accuracy assessment of classification

Class	Metrics (%)	1999		2009		2020	
		RF	SVM	RF	SVM	RF	SVM
Built-up	UA	98.1	80.9	97.8	91	97.7	54.9
	PA	99.9	92.7	93.8	98.9	100	96.6
Forest	UA	99.8	90	100	99	97.71	81.7
	PA	92.6	86.7	92.2	99.9	100	86.8
Roads	UA	99.8	81.5	92.7	66.7	99.8	88
	PA	93.02	72.1	99.8	99.9	100	77.1
Cultivable land	UA	88.2	71.3	94.6	95.3	98	82.9
	PA	98.9	84.6	94.6	87.2	90.9	90.7
Barren land	UA	99.9	85.6	95	95.1	100	77.8
	PA	99.8	56.3	88.4	91	90	54.8
OA		96.2	81.9	94.9	94.5	96.6	76.2
Kappa		0.95	0.7	0.93	0.93	0.96	0.7

particular with Sentinel-2 data, 0.4% higher than the OA of the 1999 Landsat classification, and 1.7% higher than the OA of the 2009 Landsat classification. Regarding SVM, the higher result was obtained with the Landsat image for 2009, with a significant percentage difference of 18.3% compared to the one of 2020 and 12.6% compared to the 1999 SVM results. Between the two algorithms, regardless of the type of input image, RF outperformed SVM by 20.4% for 2020, 0.4% for 2009, and 14.3% for 1999.

4.2 Analysis of LULC Changes

The LULC classification maps for the years 1999, 2009, and 2020 are illustrated in Fig. 3. Notably, LULC has undergone considerable changes between 1999 and 2020. Additionally, according to the results of area calculations for each LULC class represented in Table 2, changes in LULC trends were observed for all LULC types.

Indeed, in 1999, the Allala watershed was dominated by cultivable land area, with a percentage of 53%, followed by barren lands (28%), forest (16.1%), built-up (1.7%), and roads (0.3%).

In 2009, considerable changes were noted in LULC trends, with the amount of land devoted to roads significantly increasing (from 0.2% to 12.2%) for the total area. Similarly, the built-up class increased remarkably (Fig. 3 and Table 2). In contrast, the barren class has undergone a significant decrease (from 28.4% to 13%) for the total area.

In 2020, there was a similar decrease for the barren land class, down to 8.7% for the total area. In parallel, roads and built-up continued to increase, and there was a slight increase in cultivable lands area (from 48.8% to 49.9%). A slight decrease also was observed in forests (from 18.4% to 14%) for the total LULC area.

Over the 3 years, cultivable lands stand as the dominant class in terms of surface area as compared to other classes. In addition, during the studied period (1999–2020), the Allala watershed saw considerable changes in LULC classes, where a decrease in some classes corresponded to an increase in others. These changes can be better explained by analyzing the LULC transition matrix.

4.3 Analysis of LULC Transition Matrix

Detailed results of the LULC transition matrices from 1999 to 2009 and 2009 to 2020 are shown in Tables 3 and 4. The analysis of the LULC transition matrices indicates significant trends in LULC transitions from one LULC type to another. According to Table 3 (1999–2009), a high transition rate was observed in the forest class, which converted into cultivable lands with 773.4 ha, cultivable lands into

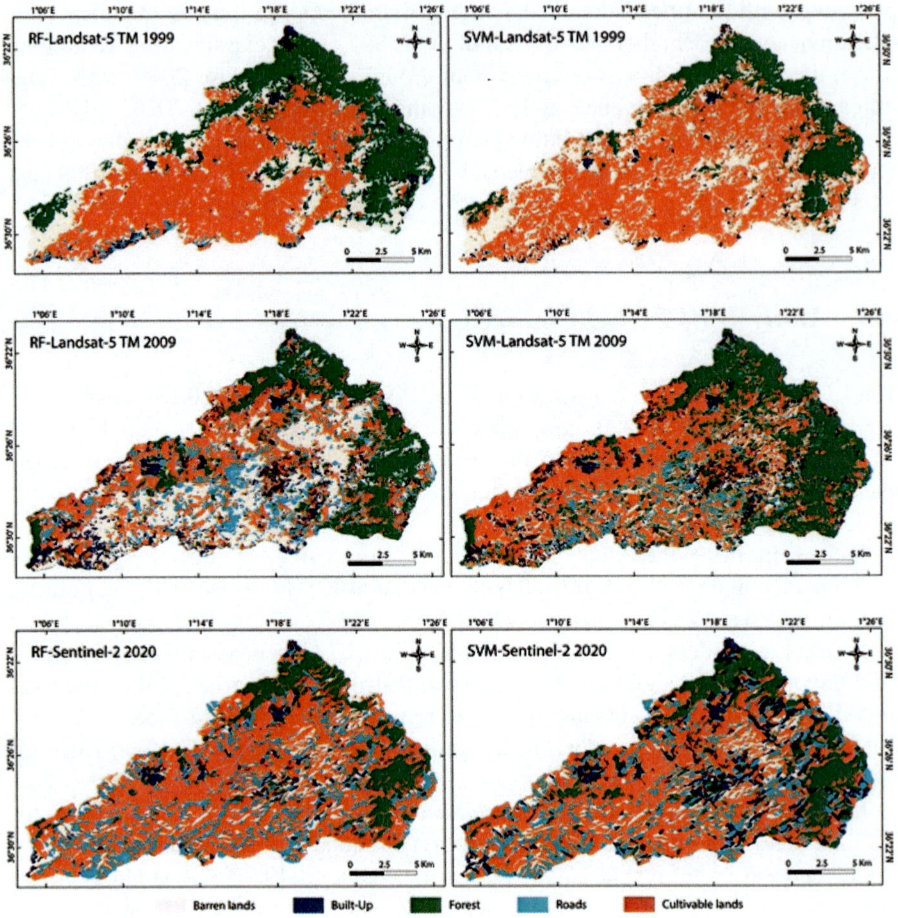

Fig. 3 LULC classification of 1999, 2009, and 2020

Table 2 LULC class areas

	1999		2009		2020	
	Area (ha)	% of total area	Area (ha)	% of total area	Area (ha)	% of total area
Built-up	537.2	1.7	2339.3	7.6	3720.1	12.1
Forest	4957.4	16.1	5652.2	18.4	4300.1	14
Roads	96.7	0.3	3736	12.2	4714.1	15.4
Cultivable land	16387.8	53.4	15,001	48.8	15307.5	49.9
Barren land	8726.6	28.4	3979	13	2664	8.7

Table 3 Transition matrix for 1999–2009 period

	LULC classes (2009) (ha)					
LULC classes (1999) (ha)	Built-up	Forest	Roads	Cultivable lands	Barren lands	Total
Built-up	359.7	15.8	30.5	34.9	92.6	533.4
Forest	14	3659	131.1	773.4	368.5	4946
Roads	7.7	7.1	10.7	27.1	43.9	96.5
Cultivable lands	373.7	1344.1	2373.7	6294.7	5989.5	16375.7
Barren lands	1577.3	611.9	1185.5	3180.7	2152.8	8708.2
Total	2332.3	5637.9	3731.6	10310.7	8647.3	30659.8

Table 4 Transition matrix for 2009–2020 period

	LULC classes 2020 (ha)					
LULC classes 2009 (ha)	Built-up	Forest	Roads	Cultivable lands	Barren land	Total
Built-up	977.5	20.3	486.7	119.0	729.8	2333.4
Forest	338.0	3186.8	176.3	1436.8	502.9	5640.8
Roads	570.9	122.6	833.3	898.0	1307.5	3732.3
Cultivable lands	1173.6	624.3	1664.9	4698.2	2153.8	10314.9
Barren land	652.9	341.7	1544.1	1896.6	4214.0	8649.3
Total	3713.0	4295.8	4705.3	9048.6	8908.0	30670.8

barren lands with 5989.5 ha, and barren lands to roads, built-up, and cultivable lands with 1185.5 ha, 1577.3 ha, and 3180.7 ha, respectively.

Regarding transitions between 2009 and 2020 (Table 4), the most significant transitions were reported for forest into cultivable lands with 1436.8 ha for the total forest area, barren land into cultivable lands with 1896.6 ha and roads with 1544.1 ha, and cultivable lands into barren lands, roads, and built-up with 2153.8 ha, 1664.9 ha, and 1173.6 ha, respectively.

5 Discussion

In this study, we aimed to map and detect LULC changes in the Allala watershed for over 21 years. Both the OBIA method based on machine learning classifiers and Arc GIS applications were used in the analysis of two Landsat images and one Sentinel-2 image. These images were gathered using remote sensing techniques and covered the years 1999, 2009, and 2020. The LULC classes in question were: built-up, forest, barren land, roads, and cultivable lands. Furthermore, in order to achieve the best possible accuracies in the change detection step, two machine learning algorithms were tested: RF and SVM. The best of both was used to detect LCLU changes.

The RF and SVM machine learning algorithms both achieved good results, with an overall accuracy ranging from 76% to 96%. In addition, the RF algorithm outperformed SVM in all LCLU classifications, regardless of the remote sensing types

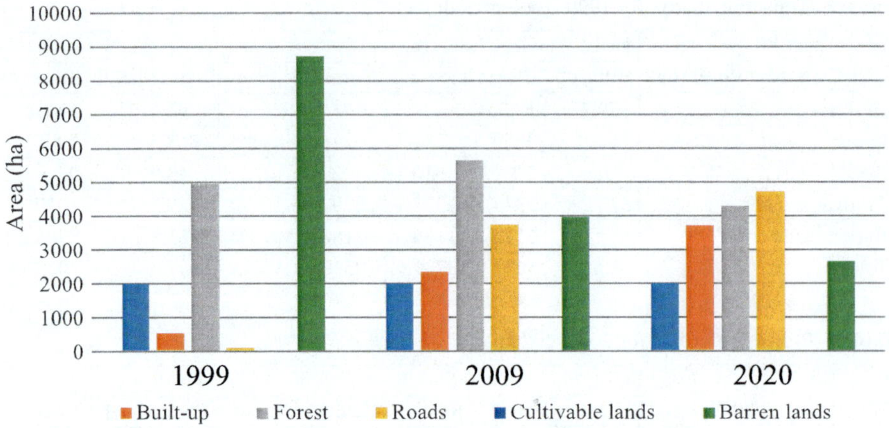

Fig. 4 Area of LULC classes of the total area (ha) over 1999, 2009, and 2020

used in the classification process. This is mentioned in several studies, for instance, [3, 15, 19, 38]. In our case, the optimization of parameters of the used classifiers has allowed us to obtain better results with high values of the overall accuracy.

Additionally, our experiments exhibited that RF performed better on the Sentinel-2 image than on the Landsat images and gives the highest value of the overall accuracy (up to 96%). These results confirm the efficacy and robustness of this algorithm when applied to this type of data, particularly for LULC mapping [39, 40]. In addition, this is strongly explained by the spatial resolution effect.

By UA and PA calculations (Table 1), the accuracy of the classes has also been evaluated within each algorithm for the 3 years. As well, for 1999 and 2020, built-up and barren land were well detected with RF, with UA and PA values at more than 96%. For 2009, forest was well classified with an UA and a PA of more than 92%. However, the worst class in terms of UA and PA was barren lands with values close to 54%. However, despite the high values of the overall precision of some classes especially roads. This class was not well delineated and took up a large area compared to other classes. This is due to the effect of the segmentation of the satellite image and in particular the "scale" parameter which generated large segments which then affected the final classification.

Considering the results of LULC change detection over the two periods, 1999–2009 and 2009–2020, as observed from the final LULC maps (Fig. 3), in addition to the results shown in Table 2 and Fig. 4, the Allala watershed has undergone different change trends over the studied years. Additionally, according to transition matrix results over 1999–2009 and 2009–2020 (Tables 3 and 4), the LULC is progressively transitioning from one LULC class to another.

Obviously, a great increase in area of both built-up and roads is observed, which is explained by the population growth and subsequent urbanization. This also facilitated the transition of some classes to built-up and roads. This increase was offset by a big decrease in the barren land class, 15.4% from 1999 to 2009 and 4.3% from

2009 to 2020. Specifically, the change rate is more significant during the 1999–2009 period. These results are logical, as Ténès City has been under constant urban construction over the last decades.

As well, the forest class has decreased by 2.1% from 1999 to 2020. This is explained by the degradation of watershed ecosystems and the many forest fires this region has experienced. Furthermore, the cultivable land class, which occupies the largest proportion of the total watershed area (53.4% in 1999), has decreased by 4.6% from 1999 to 2009 and then increased by 1.1% from 2009 to 2020. This is generally related to the harvest season when satellite images were acquired.

Moreover, forest area has been converted into barren land (7.7%) and cultivable land (15.6%). Regarding transitions between 2009 and 2020, the most important transitions were reported for forest into cultivable lands (16%) and barren land into built-up (18%), roads (14%), and cultivable lands (37%). This is explained by the forest degradation in the Allala watershed.

6 Conclusion

This study mapped and assessed land use and land cover changes in the Oued Allala watershed for over 21 years. The OBIA approach applied on two Landsat-5 TM images, taken in 1999 and 2009, and one Sentinel-2 image from 2020 enabled the identification of the changes in LULC and the distribution of area classes in the total watershed area for over 21 years. Consequently, machine learning classifiers applied with OBIA obtained higher accuracies of LULC classification (>90% overall accuracy). Specifically, the RF machine learning classifier outperformed SVM with all data over the 3 years (> 96%). In addition, the proposed methodology combined remote sensing results with Geographic Information System (GIS) tools in order to assess LULC changes efficiently.

According to the results of LULC change detection, Oued Allala has undergone many changes over time. These results confirm that the region has undergone many environmental changes in recent decades due to the growth in population, urbanization, and the evolution of agricultural areas. Notable changes include an increase in building (15.4%), cultivable land (1.1%), and road (10.9%) classes and a decrease in forests (4.4%) and barren land (4.3%). Hence, the change rate is more significant in the 1999–2009 period. These statistics are logical, as Ténès City was still under urban construction during this period. However, analyzing the transition matrix results showed that the most significant transitions were reported for forest into cultivable lands (16%), which is confirmed by the deforestation in the region. Barren land also transited into built-up (18%), roads (14%), and cultivable lands (37%).

Analyzing and mapping the trends of LULC changes in the studied area provide a basis for strategic planning, managing, and protection decision-making, and the results of LULC change analysis can be used as a decision support tool and further help in regional and national land management.

Acknowledgments This research was funded by the Hubert Curien PHC-Tassili project (19 MDU 207), the interdisciplinary graduate school for the blue planet (ISblue), and the program of ERASMUS + Chair Jean Monnet European Spatial Studies of Sea and Coastal zones – 599967-EPP-1-2018-1-FR-EPPJMO-CHAIR.

References

1. Abdullah AYM et al (2019) Spatio-temporal patterns of land use/land cover change in the heterogeneous coastal region of Bangladesh between 1990 and 2017. Remote Sens 11:790
2. Adafer S, Bensaibi M (2017) Seismic vulnerability classification of roads. Energy Procedia 139:624–630
3. Adugna T, Xu W, Fan J (2022) Comparison of random forest and support vector machine classifiers for regional land cover mapping using coarse resolution FY-3C images. Remote Sens 14:574
4. Baatz M, Schäpe A (2000) Multiresolution Segmentation: an optimization approach for high quality multi-scale image segmentation. Proceedings of Angewandte Geographische Informationsverarbeitung XII, p 12
5. Belgiu M, Csillik O (2018) Sentinel-2 cropland mapping using pixel-based and object-based time-weighted dynamic time warping analysis. Remote Sens Environ 204:509–523
6. Bello OM, Aina YA (2014) Satellite remote sensing as a tool in disaster management and sustainable development: towards a synergistic approach. Procedia Soc Behav Sci 120:365–373
7. Bentekhici N (2020) Contribution of remote sensing and GIS to mapping the fire risk of Mediterranean forest case of the forest massif of Tlemcen (North-West Algeria). Nat Hazards 21
8. Blaschke T (2010) Object based image analysis for remote sensing. ISPRS J Photogramm Remote Sens 65:2–16
9. Castillejo-González IL et al (2009) Object- and pixel-based analysis for mapping crops and their agro-environmental associated measures using QuickBird imagery. Comput Electron Agric 68:207–215
10. Chen G, Hay GJ, Carvalho LMT, Wulder MA (2012) Object-based change detection. Int J Remote Sens 33:4434–4457
11. Chen Y, Ming D, Lv X (2019) Superpixel based land cover classification of VHR satellite image combining multi-scale CNN and scale parameter estimation. Earth Sci Inform 12:341–363
12. eCognition (2021) [En ligne]. Disponible sur: https://fr.geospatial.trimble.com/products-and-solutions/ecognition
13. Farjad B, Gupta A, Razavi S, Faramarzi M, Marceau D (2017) An integrated modelling system to predict hydrological processes under climate and land-use/cover change scenarios. Water 9:767
14. Garg V et al (2019) Human-induced land use land cover change and its impact on hydrology. Hydro Res 1:48–56
15. Ghosh A, Joshi PK (2014) A comparison of selected classification algorithms for mapping bamboo patches in lower Gangetic plains using very high resolution WorldView 2 imagery. Int J Appl Earth Obs Geoinf 26:298–311
16. Guan D et al (2011) Modeling urban land use change by the integration of cellular automaton and Markov model. Ecol Model 222:3761–3772
17. Halmy MWA, Gessler PE, Hicke JA, Salem BB (2015) Land use/land cover change detection and prediction in the north-western coastal desert of Egypt using Markov-CA. Appl Geogr 63:101–112
18. Hansen MC, Defries RS, Townshend JRG, Sohlberg R (2000) Global land cover classification at 1 km spatial resolution using a classification tree approach. Int J Remote Sens 21:1331–1364

19. Htitiou A et al (2019) The performance of random forest classification based on phenological metrics derived from Sentinel-2 and Landsat 8 to map crop cover in an irrigated semi-arid region. Remote Sens Earth Syst Sci 2:208–224
20. How Jin Aik D, Ismail MH, Muharam FM (2020) Land use/land cover changes and the relationship with land surface temperature using landsat and MODIS imageries in Cameron Highlands, Malaysia. Land 9:372
21. Huang C, Davis LS, Townshend JRG (2002) An assessment of support vector machines for land cover classification. Int J Remote Sens 23:725–749
22. Kastali A et al (2021) Design flood and flood-prone areas under rating curve uncertainty: area of Vieux-Ténès. Algeria J Hydrol Eng 26:1–12
23. Lourenço P et al (2021) Assessing the performance of different OBIA software approaches for mapping invasive alien plants along roads with remote sensing data. Int J Appl Earth Obs Geoinf 95:102263
24. Mas J-F (1999) Monitoring land-cover changes: a comparison of change detection techniques. Int J Remote Sens 20:139–152
25. Myint SW, Gober P, Brazel A, Grossman-Clarke S, Weng Q (2011) Per-pixel vs. object-based classification of urban land cover extraction using high spatial resolution imagery. Remote Sens Environ 115:1145–1161
26. Platt RV, Ogra MV, Badola R, Hussain SA (2016) Conservation-induced resettlement as a driver of land cover change in India: an object-based trend analysis. Appl Geogr 69:75–86
27. Psomiadis E, Soulis K, Zoka M, Dercas N (2019) Synergistic approach of remote sensing and GIS techniques for flash-flood monitoring and damage assessment in Thessaly Plain Area, Greece. Water 11:448
28. Putri LM, Wicaksono P (2021) Mapping of land use changes in the core zone of parangtritis sand dunes using OBIA method 2015–2020. JG 13:109
29. Sehili F, Madani S (2017) The policies of seismic risk management: case of the protected areas of Dellys and Tenes Cities, Algeria. Int J Civ Eng 11
30. Singh A (1989) Review Article Digital change detection techniques using remotely-sensed data. Int J Remote Sens 10:989–1003
31. Sinha P, Kumar L (2013) Markov land cover change modeling using pairs of time-series satellite images. Photogramm Eng Remote Sens 79:1037–1051
32. Veldkamp A, Lambin EF (2001) Predicting land-use change. Agric Ecosyst Environ 85:1–6
33. Wang SW, Gebru BM, Lamchin M, Kayastha RB, Lee W-K (2020) Land use and land cover change detection and prediction in the Kathmandu District of Nepal using remote sensing and GIS. Sustainability 12:3925
34. Weng Q (2002) Land use change analysis in the Zhujiang Delta of China using satellite remote sensing, GIS and stochastic modelling. J Environ Manage 64:273–284
35. Yan G, Mas J-F, Maathuis BHP, Xiangmin Z, Van Dijk PM (2006) Comparison of pixel-based and object-oriented image classification approaches—a case study in a coal fire area, Wuda, Inner Mongolia. China Int J Remote Sens 27:4039–4055
36. Yin J, He F, Xiong YJ, Qiu GY (2017) Effects of land use/land cover and climate changes on surface runoff in a semi-humid and semi-arid transition zone in Northwest China. Hydrol Earth Syst Sci 21:183–196
37. Zaabar N, Niculescu S, Mihoubi MK (2021) Assessment of combining convolutional neural networks and object based image analysis to land cover classification using Sentinel 2 satellite imagery (Tenes region, Algeria). Int Arch Photogramm Remote Sens Spatial Inf Sci XLIII-B3-2021:383–389
38. Zaabar N, Niculescu S, Mihoubi MK (2022) Application of convolutional neural networks with object-based image analysis for land cover and land use mapping in coastal areas: a case study in Ain Témouchent, Algeria. IEEE J Sel Top Appl Earth Obs Remote Sens:1–16. https://doi.org/10.1109/JSTARS.2022.3185185

39. Zeroual A, Assani AA, Meddi M (2017) Combined analysis of temperature and rainfall variability as they relate to climate indices in Northern Algeria over the 1972–2013 period. Hydrol Res 12
40. Zope PE, Eldho TI, Jothiprakash V (2017) Hydrological impacts of land use–land cover change and detention basins on urban flood hazard: a case study of Poisar River basin, Mumbai, India. Nat Hazards 87:1267–1283

Deep Learning–Based Bathymetry Mapping from Multispectral Satellite Data Around Europa Island

Khishma Modoosoodun Nicolas, Lucas Drumetz, Sébastien Lefèvre, Dirk Tiede, Touria Bajjouk, and Jean-Christophe Burnel

Abstract Bathymetry studies are important to monitor the changes occurring in coastal topographies, to update navigation charts, and to understand the dynamics of the marine environment. Satellite-derived bathymetry enables rapid mapping of large coastal areas through measurement of optical penetration of the water column. In this study, bathymetry prediction is investigated using Pleiades multispectral satellite data. This research work explores the possibility of using very-high-resolution multispectral satellite data with a deep learning U-Net-inspired neural network architecture to infer bathymetry estimates around Europa Island (22°20′S, 40°22′E), which is a coralline island in the Mozambique Channel. This study is among the first to provide an overview suitable for bathymetry mapping using a deep learning approach based on optical satellite data. An airborne light detection and ranging (LiDAR) dataset of 1 m resolution is used as ground truth to train the model. From experiments, the overall accuracy evaluation of the model shows a good relationship

K. M. Nicolas (✉)
Université Bretagne Sud, UMR CNRS IRISA, Vannes, France

Paris Lodron University of Salzburg, Salzburg, Austria
e-mail: khishma.modoosoodun@stud.sbg.ac.at

L. Drumetz
IMT Atlantique, UMR CNRS Lab-STICC, Brest, France
e-mail: lucas.drumetz@imt-atlantique.fr

S. Lefèvre · J.-C. Burnel
Université Bretagne Sud, UMR CNRS IRISA, Vannes, France
e-mail: sebastien.lefevre@irisa.fr; jean-christophe.burnel@irisa.fr

D. Tiede
Paris Lodron University of Salzburg, Salzburg, Austria
e-mail: dirk.tiede@plus.ac.at

T. Bajjouk
Ifremer, Dynamiques des Ecosystèmes Côtiers, LEBCO, Plouzané, France
e-mail: touria.bajjouk@ifremer.fr

S. Niculescu (ed.), *European Spatial Data for Coastal and Marine Remote Sensing*, https://doi.org/10.1007/978-3-031-16213-8_6

($R^2 = 0.99$, standard error = 0.492) between the predicted and reference depth values that satisfy the International Hydrographic Organization (IHO) S-57 Category of Zone of Confidence (CATZOC) levels A1, A2, B, and C (IHO, 2014). These predicted bathymetry values could potentially be incorporated into electronic navigational charts. The image reconstruction shows accurate results to estimate bathymetry in the shallow waters with mean absolute error not exceeding 1.5 m in that case. The U-Net-inspired deep learning technique exhibits promising outcomes to predict water depth from very-high-resolution satellite data to operate bathymetry mapping automatically over a wide area.

Keywords Bathymetry mapping · Europa Island · Pleiades satellite · LiDAR · Deep learning · U-Net architecture · Remote sensing

1 Introduction

Bathymetry refers to the science of determining the depth of the water column in relation to sea level. Information on the topography of the seafloor is one of the essential parameters that is required to plan for any man-made construction or activity in the coastal zones. Detailed bathymetry data also provide essential information that can be used for habitat mapping [16] and studies on the distribution of benthic fauna [17]. Remote sensing techniques can provide repeated high-resolution bathymetry data for extensive ocean areas at low cost. These remotely sensed data are measured from remote platforms such as satellites, planes, or drones. Airborne laser bathymetry (ALB), such as light detection and ranging (LiDAR) systems, used in this study has gained popularity recently and possesses the advantages of being less time-consuming as compared to traditional methods to measure bathymetry in shallow and clear coastal waters [13]. In spite of its performance, this technique remains expensive for operational monitoring use.

Deciphering shallow-water depth by the analysis of multispectral and hyperspectral satellite images is called satellite-derived bathymetry (SDB). Since the 1970s, several studies have been undertaken using multispectral images to infer water depth [6, 14, 15, 19, 20, 22, 24, 26]. Depth information from ocean areas that were previously inaccessible by boats can be easily obtained by satellite images. These methods enable rapid mapping of large coastal areas through measurement of optical penetration of the water column. In this study, water depth estimation is investigated via a deep learning approach by training a model to learn mapping Pleiades 1B multispectral satellite data to bathymetry estimates coming from airborne LiDAR around the Europa Island. Both datasets are resampled to 1 m spatial resolution. Thus, at inference time, bathymetry estimates can be obtained using only the trained model and multispectral data. There are only a few, very recent publications that describe oceanic bathymetry mapping using different deep learning methods [1,

21, 25, 29]. This research study will contribute toward a new approach to derive SDB that could be used to update navigational charts around Europa Island. This study will further support the Indian Ocean sea Turtle (IOT) project to study the functional environmental habitat for marine turtles [12].

2 Aim and Objectives

We investigate bathymetry mapping using a U-Net-inspired deep learning approach to infer water depth estimation from Pleiades satellite imagery around Europa Island.

The main objectives of this study are as follows:

- Develop a deep learning architecture to find and learn a relationship between the optical image and the LiDAR-derived bathymetry data to estimate water depth.
- Validate optical satellite–derived bathymetry using airborne LiDAR dataset as ground truth.
- Perform evaluation for bathymetry estimation and compare with IHO standards.
- Produce bathymetry prediction maps from the deep learning U-Net-inspired model.

3 Study Area and Data Sources

The study site is the Europa Island that is a territory of France, located in the Mozambique Channel (Fig. 1). The island is situated at 22°20′S, 40°22′E and is surrounded by a fringing reef with a total area of 30 km^2. A mangrove forest covers over 700 hectares around an open lagoon in Europa Island [18].

Table 1 shows the data sources used in this study. A Pleiades 1B multispectral optical satellite image is used as input image for estimating bathymetry. The Pleiades image was acquired on 29 July 2013 with no cloud coverage. The image was pansharpened and resampled to 1 m spatial resolution. The bathymetric data was acquired on 21 February 2012 from an airborne topo-bathymetry LiDAR survey and was carried out in the framework of Litto3D program. The LiDAR data has a very high spatial resolution of 1 m and allows deriving a Digital Terrain Model (WGS84 georeferenced DTM) of both land and sea. For the whole island, the LiDAR DTM value ranges from −40 m below to 16.8 m above sea level. The LiDAR data was randomly split into a training and a validation dataset in the processing phase. It should be noted that the LiDAR image used for the study has been processed for tidal corrections by SHOM (Service Hydrographique et Océanographique de la Marine) according to IHO standards.

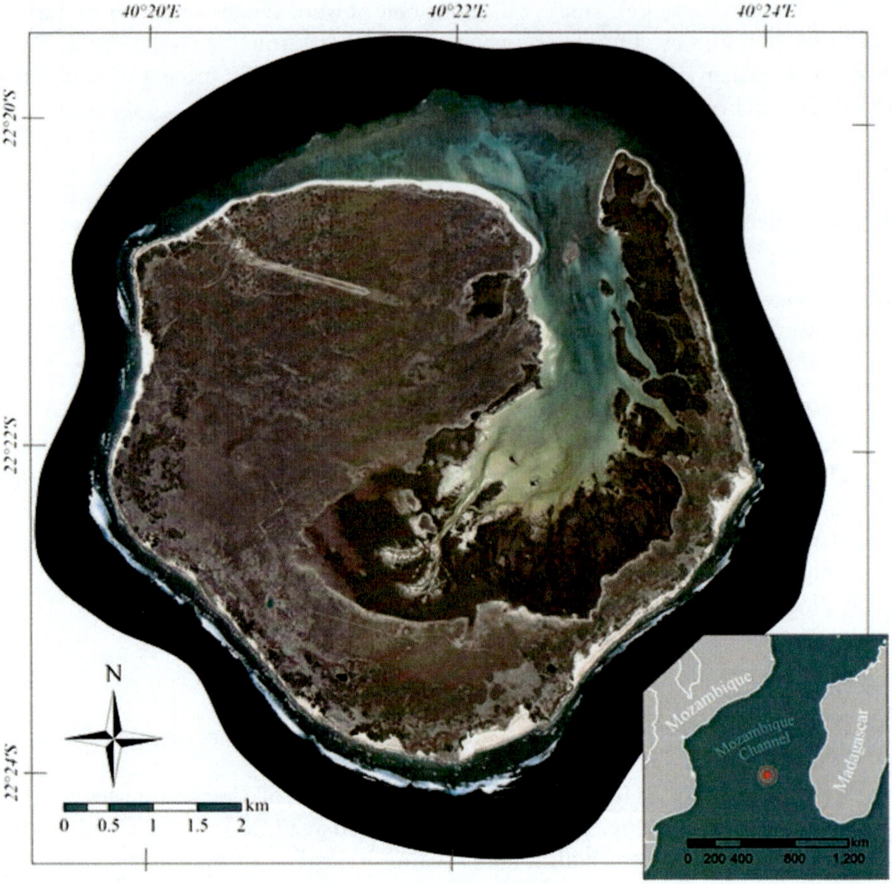

Fig. 1 Map of Europa Island, Mozambique Channel (true color Pleiades satellite image pansharpened and resampled to 1 m resolution)

Table 1 Satellite and airborne data sources

Data type	Acquisition date	No. of spectral bands (μm)	Resolution	Dimension (pixels)
Pleiades 1B Satellite Imagery (multispectral)	2013-07-29	4 Blue (0.43–0.55) Green (0.50–0.62) Red (0.59–0.71) NIR (0.74–0.94)	2 m	5033 × 5194
Pleiades 1B Satellite Imagery (Panchromatic)	2013-07-29	1 (0.47–0.83)	0.5 m	20128 × 20775
Lidat airborne data	2012-02-21	1	1 m	9000 × 10000

4 Methodology

To solve bathymetry mapping automatically and at a higher level of accuracy, deep learning methods have recently been explored. The prediction of depth maps from RGB images using deep learning, a related task, was performed by Eigen et al. [7]. This concept has been used as a reference in our study whereby a convolutional neural network and regression method were used to produce the bathymetry predictions. A U-Net-inspired convolutional neural network architecture is used to address the research question of bathymetry estimation from space. The proposed U-Net model uses a single satellite image as input and one ground truth single image, both divided into batches of input/output image patches. We rely on an existing network tailored for depth map prediction from RGB single image using NYU Depth V2 dataset [5] that comes with a publicly available GitHub repository [9]. We then adapt this existing algorithm to the specific problem of bathymetry estimation. While our task is related to U-Net semantic segmentation, a depth value is assigned to each pixel of the image to predict bathymetry, leading us to solve a regression problem instead of a classification one. To fine-tune the network, several experiments were run to evaluate the best result where the hyperparameters, activation functions, and optimizers were set. The deep learning U-Net-inspired model was implemented using Python programming with the PyTorch framework accelerated using graphics processing units (GPU).

The overall research pipeline used in this study is shown in Fig. 2. Pre-processing of the datasets (image calibration and pansharpening) has been conducted before training our model.

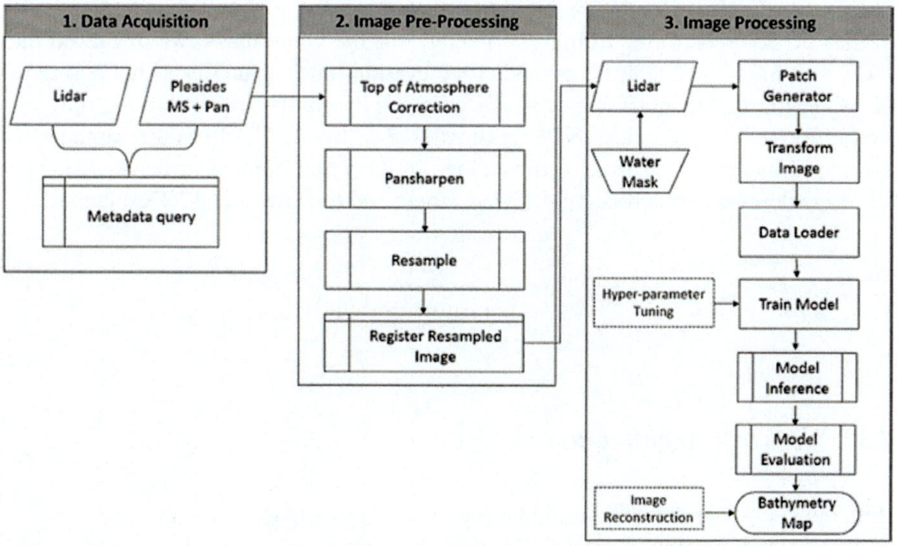

Fig. 2 Workflow of the bathymetry prediction from pre-processing to training and testing the U-Net-like model

4.1 Image Pre-processing

The input data that is to be fed into the deep learning network is the multispectral optical imagery with four spectral bands (red, blue, green, and near infrared). The pre-processing steps are performed in ArcGIS Pro version 2.7.

The multispectral data is calibrated into top of atmosphere (ToA) reflectance values using an apparent reflectance function followed by a pansharpening process. This is done to reduce any variations so as to yield accurate and reliable water depths. The apparent reflectance function calibrates the digital number (DN) values in the Pleiades imagery by adjusting reflectance or brightness in the image by using the sun elevation, acquisition date, sensor gain and bias, and sun angle correction for each band to derive the ToA reflectance.

A pansharpening process is carried out to improve the spatial resolution of the Pleiades satellite image that is used as input image for the study [28]. The pansharpened image is obtained by a fusion of the very-high-resolution panchromatic image (0.5 m) with the lower-resolution multiband Pleiades image (2 m). The computation is performed in ArcGIS Pro by using the Gram–Schmidt algorithm [23]. Ultimately, this process ensures that the deep neural network better extracts the spatial features from the Pleiades image.

In order to feed the input image in the deep learning network, it should have the same dimension and spatial resolution as the ground truth (LiDAR) dataset. A resampling of the input Pleiades image is done to match the same image dimension of the LiDAR data (10,000 by 9000 pixels and spatial resolution of 1 m). As resampling technique, a bilinear interpolation, useful for continuous data values, is performed.

Since we are interested in bathymetry, we need to discard land areas in the image so they do not perturb the training. To determine the water mask, we threshold the DTM to 3 m and consider all pixels with elevation lower than 3 m as belonging to the water mask. The mask was visually assessed by comparing it with the normalized difference water index (NDWI) illustrated in Eq. 1. NDWI is a remote sensing–based indicator sensitive to the change in the water content of leaves [8]. The NDWI is defined as the relative difference between the Green and NIR channels.

$$\mathrm{NDWI} = \frac{(\mathrm{Xgreen} - \mathrm{Xnir})}{(\mathrm{Xgreen} + \mathrm{Xnir})} \tag{1}$$

4.2 Network Architecture

Both the LiDAR and the resampled multispectral image have the same spatial resolution of 1 m and a dimension of 10,000 by 9000 pixels. A patch generator is implemented to divide the images into 64 by 64 small water-only input/output pairs of patches, based on the derived water mask. They are transformed using a min–max scaler normalization to ensure stable training, before being fed to the network that processes batches of multispectral/LiDAR patches.

The U-Net network architecture (Fig. 3) was chosen as a base model to perform a semantic regression approach as it helps in better prediction of pixel-wise water depth value. U-Net architecture and variants became popular for the semantic segmentation task. This is due to their fully convolutional nature and their capacity to pass through information at different spatial scales, thanks to skip connections, added on top of a convolutional encoder/decoder architecture. The encoder uses pooling operations to progressively reduce the dimensionality of the features, while the decoder uses transposed convolution layers to retrieve an image with the same dimensions as the input. The training ran for 50 epochs with batch size of 128, an ADAM optimizer, and ReLU activation function at a learning rate of 0.001. In this research work, a scale invariant error loss function (Eq. 2) as proposed by Eigen et al. [7] is used in training the model. In Eq. 2, y^* denotes the predicted water depth, y is the ground truth image, n is the number of pixels, d_i is the difference between the absolute values and the ground truth, and $\lambda = 0.5$ a regularization parameter set similar to Eigen et al. [7].

$$L\left(y,y^*\right) = \frac{1}{n}\sum_i d_i^2 - \frac{\lambda}{n^2}\left(\sum_i d_i\right)^2 \tag{2}$$

Learning is performed to find the best combination of model hyperparameters that aims at minimizing the loss function from the training optical image and its related target (ground-truth DTM). This is done by drawing random batches of data samples and their associated targets to subsequently compute the gradients of the loss

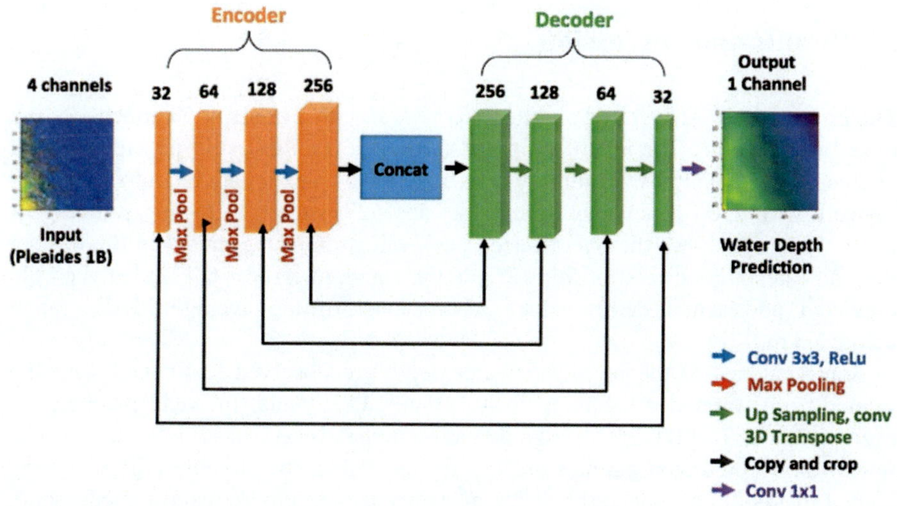

Fig. 3 Proposed U-Net-inspired architecture for bathymetry prediction

with respect to the network parameters on the chosen batch size (thanks to automatic differentiation provided by PyTorch). Eventually, the U-Net model inference is evaluated to assess the bathymetry predictions.

4.3 Model Evaluation

One of the main challenges for solving depth regression problems is defining an appropriate loss function. Following the various pixel-wise regression research studies for depth estimation that have shown promising results using standard regression loss [2], we chose to rely on the mean absolute (L1) (Eq. 3) and root-mean-square loss (L2) (Eq. 4). The mean absolute error (MAE) (L1) is the commonly used loss for depth estimation that is directly interpretable in terms of distance [2]. These water depth predictions produced from the U-Net network are evaluated as mean absolute error and root mean squared error (RMSE) values to determine how far the predictions matched with the ground truth. The MAE and RMSE are calculated for individual patches and globally for all patches of the training and testing datasets as described below.

$$MAE = \frac{1}{n}\sum_{j=1}^{n}\left|y_j - \hat{y}_j\right| \tag{3}$$

$$\text{RMSE} = \sqrt{\frac{1}{n}\sum_{j=1}^{n}\left(y_j - \hat{y}_j\right)^2} \tag{4}$$

5 Results and Discussion

The proposed U-Net-inspired architecture took around 2 hours at 50 epochs for the model to converge. The result of an 80/20 ratio of train/test split provided a total number of 2632 samples. Randomly chosen, 64 by 64 patches are fed into the U-Net network with 2105 patches as training set and 527 patches as testing set, respectively. Figure 4 shows the overall accuracy evaluation of the model in the test set that indicates a good relationship ($R^2 = 0.99$, standard error = 0.492) between the predicted and actual depth values where most of the average MAEs range below 1.5 m.

Some patches of the predicted water depth are observed to fit well with the ground-truth water depth having low-average MAE value. In some patches, for example, patch ID 60 (Fig. 5), high deviation in the average MAE is observed. The few wrongly predicted patches are mostly located in the shallow regions of the lagoon that could be due to the differences in substrate types (seagrass beds, sand,

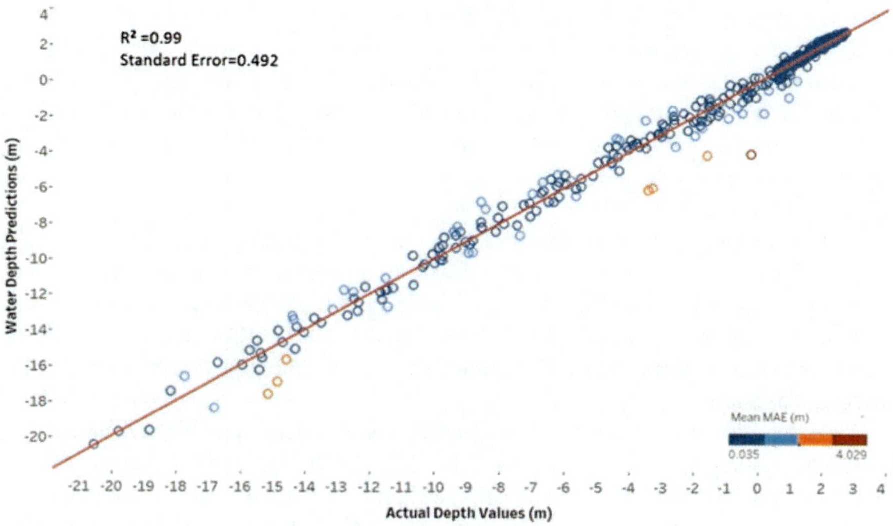

Fig. 4 Scatterplot of predicted values against actual values

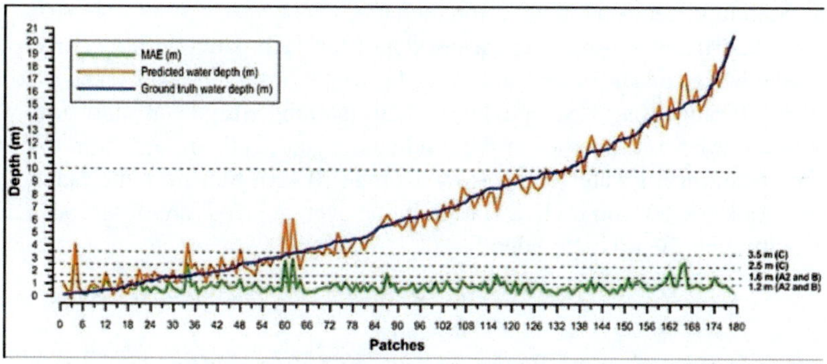

Fig. 5 Accuracy comparison according to IHO S-57 CATZOC level for the predicted and actual depth at Europa Island

and corals). Different substrates have different reflectance values, and this could influence the depth prediction in these shallow waters. Similar findings in shallow coastal areas have also been reported to have low accuracies [3, 27, 30, 31].

5.1 *International Hydrographic Organization (IHO) Standards*

The International Hydrographic Organization (IHO) standards can be used to evaluate the accuracy of bathymetric surveys. The IHO S-57 standard puts forward Category of Zone of Confidence (CATZOC) levels. This is used to define depth accuracy specifications for specific depth ranges, which are incorporated into electronic navigational charts to provide an indication of navigational safety to mariners, especially in shallow areas [10, 11]. The CATZOC indicator considers the total vertical uncertainty and total propagated uncertainty of the hydrographic data, together with the survey's seafloor coverage. Then, these data are merged with an algorithm to organize bathymetric values in one of the five categories, namely, A1, A2, B, C, D, and U [10, 11].

Table 2 provides an overview of the accuracy requirements (MAE) for each CATZOC level for depths up to 30 m. Each CATZOC is integrated in Electronic Chart and Display Information System (ECDIS) and represented as a series of stars in the Electronic Navigational Charts. The IHO S-57 CATZOC standards have been used in previous studies to check whether the estimated depths can be included in navigational charts [4, 29].

From the experiments, most of the predicted water depth patches fall under the IHO S-57 Category of Zone of Confidence (CATZOC) levels A1, A2, B, and C (Fig. 5). It was observed that 38.4% of patches fall in A1 standard, 51.4% meet the A2 and B standard, and only 10.2% fall in the C level (Table 2). Only one patch with a MAE of 4.029 m did not meet the CATZOC standard, which might be located in shallow or turbid waters. These predicted bathymetry values could potentially be incorporated into electronic navigational charts.

The ground-truth and predicted water depth for each patch with its corresponding average MAE are illustrated in Fig. 5. From the observed results, the predicted water depth (orange line) is seen to be fluctuating along with the ground-truth water depth (blue line). Some patches of the predicted water depth are observed to fit well with the ground-truth water depth (low-average MAE), but for some patches, for example, ID 3, 33, 60, and high deviation in the average MAE are observed, leading to overestimating the predicted depth.

5.2 *Reconstruction of Bathymetry Prediction Map*

With our U-Net-inspired deep learning method, the global spatial structure of the bathymetries around Europa Island is seen to be well preserved while performing the reconstructed prediction map as shown in Fig. 6 and Fig. 7. We note low MAE values in the shallow waters and this gets higher in deeper waters (Fig. 8), thus confirming visually that our U-Net-like method exhibits reliable overall results with respect to Table 2.

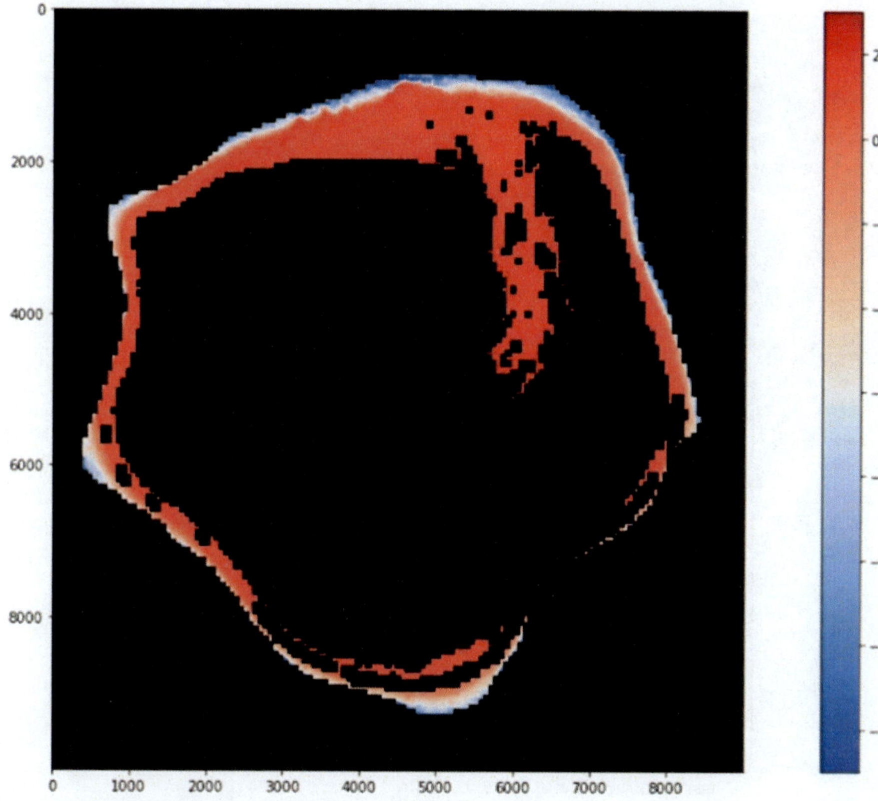

Fig. 6 Predicted bathymetry map around Europa Island

Table 2 Depth ranges for IHO CATZOC levels

CATZOC level	Depth range (m)	Required accuracy (± m)	Number of patches	(%)
A1	0–10	0.6	48	38.4
	10–30	0.8	20	
A2 and B	0–10	1.2	69	51.4
	10–30	1.6	22	
C	0–10	2.5	13	10.2
	10–30	3.5	5	

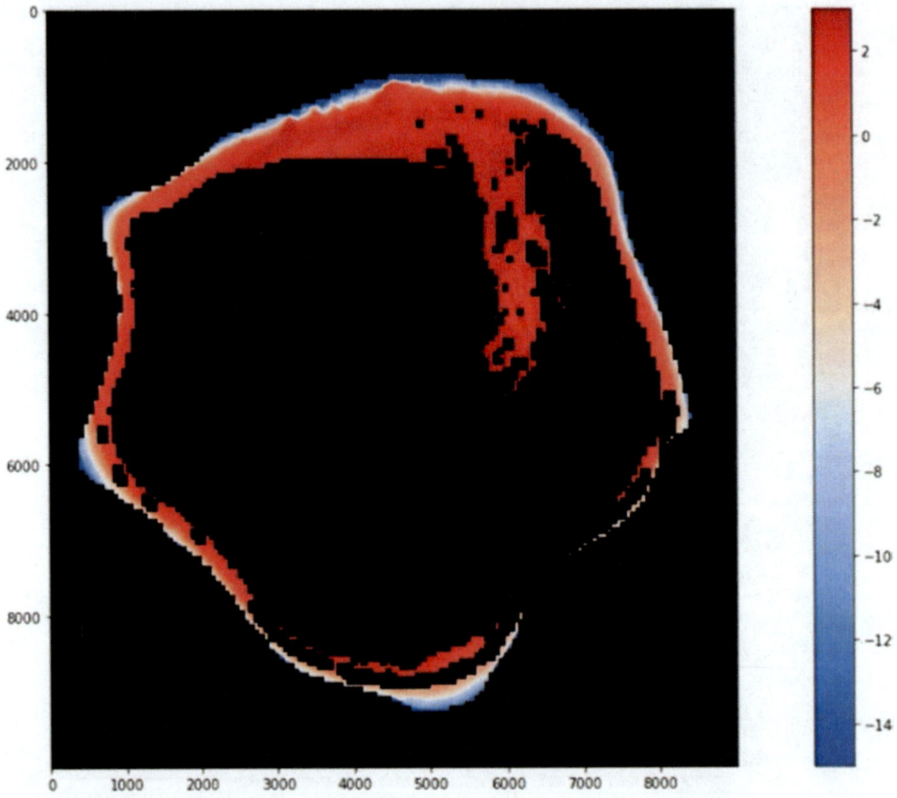

Fig. 7 Ground-truth bathymetry map around Europa Island

6 Conclusion

In this research work, a U-Net-like deep learning technique is used to estimate bathymetry around Europa Island from a multispectral Pleiades image and trained using LiDAR data. The prediction of bathymetry values from a single image was a challenging task, but it has been accomplished successfully using our U-Net-inspired model.

Our model achieves positive results to predict bathymetry using the multispectral optical data. Except for one, all the predicted bathymetry patches satisfy the IHO S-57 CATZOC standards. This shows promising results that once replicated at a larger scale and generalized to different datasets, our model could potentially be used in electronic navigational charts for Europa Island. In order to evaluate the reliability of the proposed architecture, the mean absolute error (MAE) and root-mean-squared error (RMSE) were used to assess the bathymetry predictions. It can be observed that the predicted values compared to the actual values show a good relationship. An overall accuracy evaluation of the model shows a good relationship

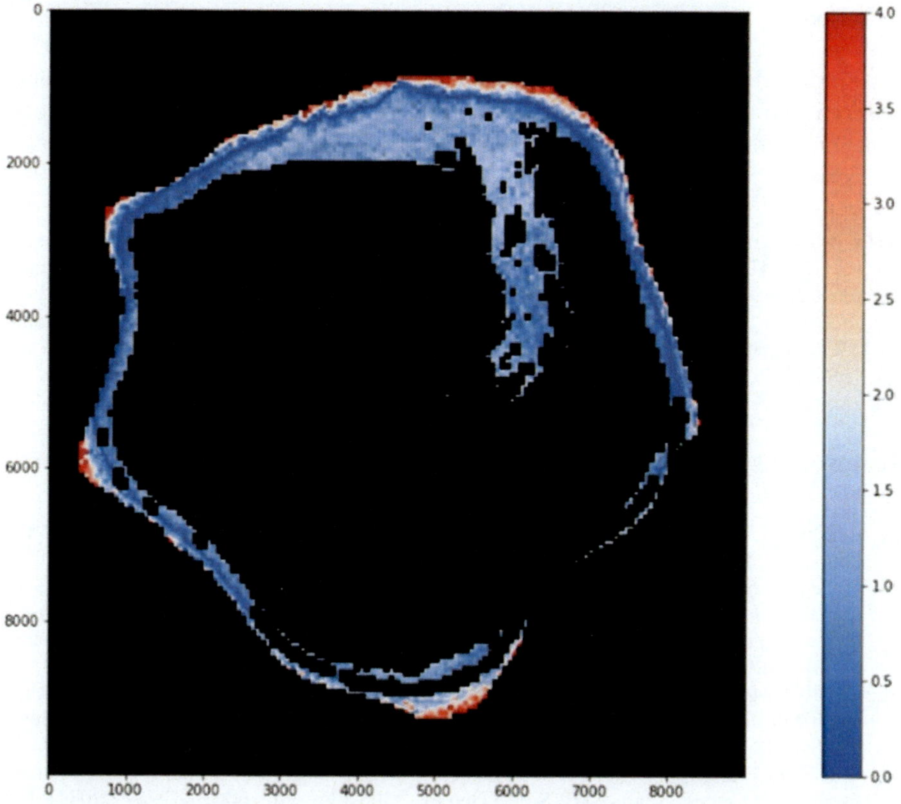

Fig. 8 Bathymetry mean absolute error map around Europa Island

(R^2 = 0.99, standard error = 0.492). Most of the average MAE values range below 1.5 m. Nearly 38.4% of patches fall in A1 standard, 51.4% meet the A2 and B standard, and only 10.2% fall in the C CATZOC level, which mean that the predicted bathymetry values can potentially be incorporated into electronic navigational charts.

To sum up, this research work is addressing bathymetry mapping using a deep learning approach, which is beneficial to provide depth estimation in remotely inaccessible areas. Such information can provide bathymetry information for a wide area in the ocean to better understand the topography of the seabed. Bathymetry studies are important to understand the dynamics of the marine environment, which is a crucial component in terms of both sediment transport and distribution of marine species. It is imperative to monitor the changes occurring in the coastal topographies in order to update navigation charts, validate hydrodynamics models, and undertake long-term wave forecasting to limit damages and save lives during extreme events such as coastal flooding. Although this technique has shown promising results to predict water depth, improvement on the model needs to be undertaken as future works.

Among possible directions for improving our model, we can mention the use of data augmentation techniques before training the model by changing the scale, rotating or flipping the images in order to increase the number of samples, and observing how the model behaves. Moreover, we can consider more advanced loss functions. Finally, it would be interesting to assess our model on other geographical areas.

Acknowledgments The first author would like to acknowledge the Erasmus Mundus Joint Master Degree (EMJMD) consortium for funding the Copernicus Master in Digital Earth joint program with the University of Salzburg and the University of South Brittany. Authors also thank the Indian Ocean sea Turtle (IOT) INTERREG project and Litto3D program for providing the satellite and LiDAR data, respectively.

References

1. Benshila R, Thoumyre G, Najar MA, Abessolo G, Almar R, Bergsma E, Hugonnard G, Labracherie L, Lavie B, Ragonneau T, Simon E (2020) A deep learning approach for estimation of the nearshore bathymetry. J Coast Res 95:1011–1015. https://doi.org/10.2112/SI95-197.1
2. Carvalho M, Le Saux B, Trouvé-Peloux P, Almansa A, Champagnat F (2018) On regression losses for deep depth estimation. In: IEEE International Conference on Image Processing (ICIP). pp 2915–2919. https://doi.org/10.1109/ICIP.2018.8451312
3. Casal G, Monteys X, Hedley J, Harris P, Cahalane C, Mccarthy T (2019) Assessment of empirical algorithms for bathymetry extraction using Sentinel-2 data. Int J Remote Sens 40:2855–2879. https://doi.org/10.1080/01431161.2018.1533660
4. Chénier R, Faucher MA, Ahola R (2018) Satellite-derived bathymetry for improving Canadian hydrographic service charts. ISPRS Int J Geo Inf 7:306. https://doi.org/10.3390/ijgi7080306
5. Couprie C, Farabet C, Najman L, Lecun Y (2013) Indoor semantic segmentation using depth information. arXiv preprint arXiv:1301.3572. https://doi.org/10.48550/arXiv.1301.3572
6. Dekker AG, Phinn SR, Anstee J, Bissett P, Brando VE, Casey B, Fearns P, Hedley J, Klonowski W, Lee ZP (2011) Intercomparison of shallow water bathymetry, hydro-optics, and benthos mapping techniques in Australian and Caribbean coastal environments. Limnol Oceanogr Methods 9:396–425. https://doi.org/10.4319/lom.2011.9.396
7. Eigen D, Puhrsch C, Fergus R (2014) Depth map prediction from a single image using a multi-scale deep network. Adv Neural Inform Proc Syst:27
8. Gao BC (1996) NDWI-A normalized difference water index for remote sensing of vegetation liquid water from space. Remote Sens Environ 58:257–266
9. GitHub repository, https://github.com/DikshaMeghwal/unet-depth-prediction. Accessed 22.02.2021
10. IHO (2014) International hydrographic organization, S-57 supplement no. 3—supplementary information for the encoding of S-57 edition 3.1 ENC data. In: Organization IH (ed) Monaco
11. IHO (2020) IHO: S-44 (2020) Standards for hydrographic surveys. Standard 6th ed., International Hydrographic Organization, Monaco. In: Organization IH (ed.). Monaco
12. IOT (2018) Pilot project for Indian Ocean sea Turtles. https://wwz.ifremer.fr/lareunion_eng/Projects/Technological-innovations/pIOT-2018-2020-IOT-2018-2021/IOT-2018-2021. Accessed 06.06.2021
13. Irish JL, Lillycrop WJ (1999) Scanning laser mapping of the coastal zone: the SHOALS system. ISPRS J Photogramm Remote Sens 54:123–129. https://doi.org/10.1016/S0924-2716(99)00003-9

14. Kao HM, Ren H, Lee CS, Chang CP, Yen JY, Lin TH (2009) Determination of shallow water depth using optical satellite images. Int J Remote Sens 30:6241–6260. https://doi.org/10.1080/01431160902842367
15. Kimeli A, Thoya P, Ngisiang'e N, Ong'anda H, Magori C (2018) Satellite-derived bathymetry: a case study of Mombasa Port Channel and its approaches, Kenya. Western Indian Ocean J Marine Sci 17:93–102. https://doi.org/10.4314/wiojms.v17i2.8
16. Kostylev VE, Todd BJ, Fader GB, Courtney R, Cameron GD, Pickrill RA (2001) Benthic habitat mapping on the Scotian Shelf based on multibeam bathymetry, surficial geology and sea floor photographs. Mar Ecol Prog Ser 219:121–137. https://doi.org/10.3354/meps219121
17. Kostylev VE, Courtney RC, Robert G, Todd BJ (2003) Stock evaluation of giant scallop (Placopecten magellanicus) using high-resolution acoustics for seabed mapping. Fish Res 60:479–492. https://doi.org/10.1016/S0165-7836(02)00100-5
18. Lambs L, Mangion P, Mougin E, Fromard F (2016) Water cycle and salinity dynamics in the mangrove forests of Europa and Juan de Nova Islands, Southwest Indian Ocean. Rapid Commun Mass Spectrom 30:311–320. https://doi.org/10.1002/rcm.7435
19. Lee Z, Carder KL, Mobley CD, Steward RG, Patch JS (1999) Hyperspectral remote sensing for shallow waters: 2. Deriving bottom depths and water properties by optimization. Appl Opt 38:3831–3843. https://doi.org/10.1364/AO.38.003831
20. Lyzenga DR, Polcyn FC (1979) Techniques for the extraction of water depth information from landsat digital data. Env Res Inst Michigan Ann Arbor Appl Div
21. Ma Y, Xu N, Liu Z, Yang B, Yang F, Wang XH, Li S (2020) Satellite-derived bathymetry using the ICESat-2 lidar and Sentinel-2 imagery datasets. Remote Sens Environ 250:112047. https://doi.org/10.1016/j.rse.2020.112047
22. Manessa MDM, Kanno A, Sekine M, Haidar M, Yamamoto K, Imai T, Higuchi T (2016) Satellite-derived bathymetry using random forest algorithm and worldview-2 imagery. Geoplanning J Geomatics Plan 3:117–126. https://doi.org/10.14710/geoplanning.3.2.117-126
23. Maurer T (2013) How to pan-sharpen images using the gram-schmidt pan-sharpen method a recipe. Int Arch Photogrammetry, Remote Sensing Spatial Inform Sci:239–244. https://doi.org/10.5194/isprsarchives-XL-1-W1-239-2013
24. Sagawa T, Yamashita Y, Okumura T, Yamanokuchi T (2019) Satellite derived bathymetry using machine learning and multi-temporal satellite images. Remote Sens 11:1155. https://doi.org/10.3390/rs11101155
25. Sonogashira M, Shonai M, Iiyama M (2020) High-resolution bathymetry by deep-learning-based image superresolution. PLoS One 15(7):e0235487. https://doi.org/10.1371/journal.pone.0235487
26. Stumpf RP, Holderied K, Sinclair M (2003) Determination of water depth with high-resolution satellite imagery over variable bottom types. Limnol Oceanogr 48:547–556. https://doi.org/10.4319/lo.2003.48.1_part_2.0547
27. Vahtmäe E, Kutser T (2016) Airborne mapping of shallow water bathymetry in the optically complex waters of the Baltic Sea. J Appl Remote Sens 10:025012. https://doi.org/10.1117/1.JRS.10.025012
28. Vivone G, Alparone L, Chanussot J, Dalla Mura M, Garzelli A, Licciardi GA, Restaino R, Wald L (2014) A critical comparison among pansharpening algorithms. IEEE Trans Geosci Remote Sens 53(5):2565–2586. https://doi.org/10.1109/TGRS.2014.2361734
29. Wan J, Ma Y (2021) Shallow water bathymetry mapping of Xinji Island based on multi-spectral satellite image using deep learning. J Ind Soc Remote Sensing 49:1–14. https://doi.org/10.1007/s12524-020-01255-9
30. Zhang A, Lipton ZC, Li M, Smola AJ (2020) Dive into deep learning. Online book (https://d2l.ai). Accessed 22.02.21. https://doi.org/10.48550/arXiv.2106.11342
31. Zhang X, Ma Y, Zhang J (2020) Shallow water bathymetry based on inherent optical properties using high spatial resolution multispectral imagery. Remote Sens 12:3027. https://doi.org/10.3390/rs12183027

Assessment of Coastal Vulnerability to Erosion Risk Using Geospatial and Remote Sensing Methods (Case of Jerba Island, Tunisia)

Amina Boussetta, Simona Niculescu, Soumia Bengoufa, Hajer Mejri, and Mohamed Faouzi Zagrarni

Abstract Located in the southeast of Tunisia, Jerba is considered a premier tourist destination offering beautiful sandy beaches. Since 1960, the island has undergone significant socio-economic transformations due to its tourism boom. Beach tourism is extremely popular among the population, which has contributed to an intensification of coastal vulnerability, wherein the beaches are threatened with disappearance. This work aims to identify the causes of coastal vulnerability and measure it based on different geophysical and socio-economic variables using the coastal vulnerability index (CVI) developed by Gornitz (Vulnerability of the US to future sea level rise, Coastal Zone, Proceedings of the 7th Symposium on Coastal and Ocean Management, American Society of Civil Engineers, 1345–1359, 1991). This allows us to identify the most vulnerable sites and to establish maps and data for coastal management purposes. The results obtained show 63% (14 km) of the coastline of the northeast coast of the island has a high to very high degree of vulnerability. Moreover, 37% of the coastline of the southeast of the island has a low to moderate

---

A. Boussetta (✉) · M. F. Zagrarni
U.R Applied Hydrosciences UR13ES81, Higher Institute of Water Sciences and Techniques, University of Gabes, Cité Erryadh, Gabes, Tunisia

S. Niculescu
Laboratory LETG-Brest, UMR 6554 CNRS, University of Western Brittany, Plouzané, France
e-mail: simona.niculescu@univ-brest.fr

S. Bengoufa
University of Western Brittany, CNRS, LETG Brest UMR 6554 CNRS, Technopôle Brest-Iroise, Av. Dumont d' Urville, Plouzané-Brest, France

National School of Marine Sciences and Coastal Management (ENSSMAL), University Campus of Dely Ibrahim Bois des Cars, Algiers, Algeria

H. Mejri
U.R Applied Hydrosciences UR13ES81, National Engineering School of Gafsa (ENIGA), University of Gafsa, Gafsa, Tunisia

S. Niculescu (ed.), *European Spatial Data for Coastal and Marine Remote Sensing*, https://doi.org/10.1007/978-3-031-16213-8_7

vulnerability or about 22 km of the entire coastline. The quantitative measures relating to this coastal vulnerability can aid to fortify the coast against a rise in sea level.

Keywords CVI · Erosion · Island · Jerba · OBIA · Remote sensing · RF · Vulnerability

1 Introduction

An unquestionable consequence of global warming is the alarming sea-level rise (SLR). In fact, the International Panel on Climate Change [29] predicts that SLR will be 0.63–1.01 m by 2100, which would be twice as fast as that which has occurred over the last 100 years. In the Mediterranean basin, the Gulf of Gabes in the southeast of Tunisia is one of the areas among the most affected by SLR, accelerating in this area much faster than the global average, estimated locally at 5.7 mm/year [24].

The concentration of population and economic activities around the coast makes coastal erosion a global concern [3]. The Tunisian coastal area makes up nearly a quarter of the territory and contains four-fifths of the urban population and a third of the rural population of Tunisia, whose total population is estimated at 11.82 million [27]. This area would face major socio-economic and environmental difficulties if no assessment of the vulnerability of coastal areas and potential adaptation measures are undertaken in the years to come. Among the degraded shorelines are the coasts of the Gulf of Gabes, with its 750 km of shoreline, including the islands of Jerba and Kerkennah, which represents 58% of the Tunisian shoreline [22]. With an area of 514 km^2, the island of Jerba is densely populated with a population of 163, 726 [27]. Jerba started attracting tourists in the 1960s and has become one of the most popular destinations in the region. The rapid development of tourist facilities (more than 130 hotels) and the increase in population in these coastal areas have led to several coastal problems.

The coastline is a highly complicated and fragile environment [20]. It is in perpetual and rapid change [6, 31], and therefore, natural dynamics, extreme weather hazards, and anthropogenic factors make coastal areas particularly vulnerable. Coastal pressure induces complications and the disturbance of coastal areas [8]. This situation has motivated research on the concept of vulnerability in order to make it measurable. Therefore, the coastal vulnerability index (CVI) calculation was developed by Gornitz [9] and has subsequently been improved at different scales by numerous studies [1]. In the present study, three geological and two physical variables and a socio-economic factor are used.

The aim of this study is to apply multiple approaches in order to assess the coastal vulnerability of Jerba Island using remote sensing methods and the Geographic Information System (GIS). The availability of remotely sensed data is emerging as one of the best means for terrestrial monitoring [14]. Starting with a Landsat-5 image using the Thematic Mapper (TM) sensor (resolution equal to 30 m), the Copernicus platform development and Theia then offered new open-access Sentinel-2 optical images, with improved spatial resolution (10 m) and satellite revisit time (5 days). However, although Sentinel-2 images offer better characteristics, very little research has been done to evaluate their potential in detecting and monitoring the shoreline to measure coastal vulnerability. Spatial remote sensing, especially the extraction of geographical objects by automatic classification methods, is becoming an important global scientific tool applied in the broad field of environmental sciences [30].

The aim of this study is to identify the causes of coastal vulnerability, measure them on the basis of different geophysical and socio-economic variables from the calculation of CVI, and identify the most vulnerable sites in the studied area. The integrated methods will build databases on coastal areas and their dynamic processes that will help managers in planning, decision-making, and development of coastal cities.

2 Study Area

Jerba, located in the Gulf of Gabes in southeastern Tunisia, is the largest Tunisian island on the North African coast, located 2 km from the mainland, 58 km east-southeast of Gabes, and 23 km north-northwest of Zarzis, at a northern latitude between 33°57′0.56" and 33°37'46.82" and an eastern longitude between 10°45′38.43" and 11°3'53.802". It covers an assembly of three maps at the scale of 1/50.000, with a surface area of 514 km^2 and boasts a 150 km shoreline.

The island is 30 km long and 20 km wide and covers 50,000 ha, plus 3000 ha of marshes and intertidal zones.

Along the southwestern side, the island is linked to the mainland between Ajim and the peninsula of Djorf by a ferry, and on the southeast side, it connects to the peninsula of Zarzis by the Roman road El Kantara, 7.5 km long (Fig. 1).

3 Methodology

Our approach focuses on a calculation of the coastal vulnerability index (CVI) by integrating variables adapted to the microtidal coasts of the western Mediterranean basin. In this study, the adopted formulation of the CVI includes six geophysical and socio-economic variables: shoreline change (a), geomorphology (b), slope (c), wave height (d), mean tidal range (e), and land use (f). These data were collected from a combination of sources, including field surveys, satellite images, topographic maps, a Shuttle Radar Topographic Mission, and various research papers.

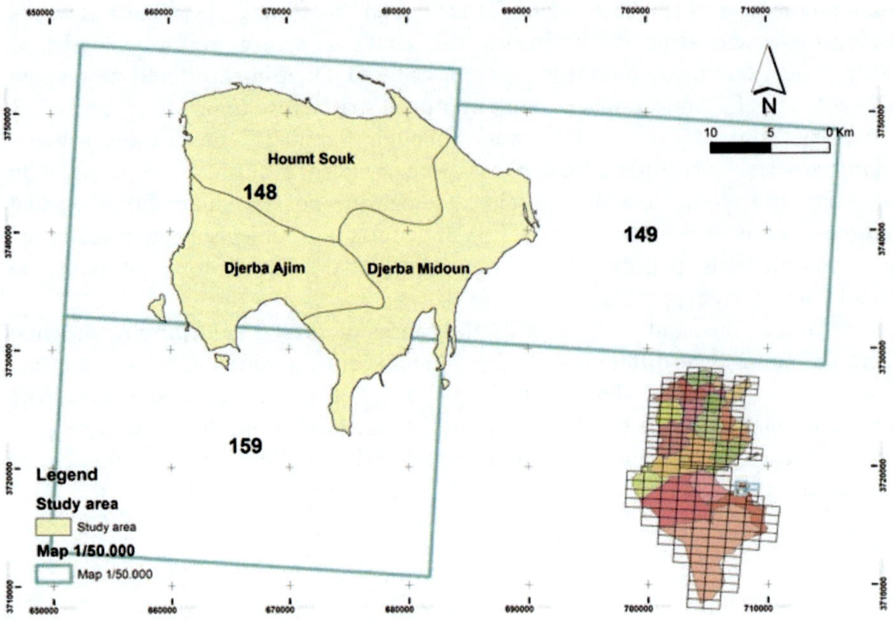

Fig. 1 Cartographic assembly covering the study area

3.1 Treatment

The analysis of the Landsat and Sentinel satellite imagery provided a variety of data that were used to process the variables chosen for the CVI model in a socio-ecosystemic environment. The variables were integrated into a GIS to calculate the CVI, after being ranked according to their impact on the coastal environment. Therefore, the entire coastline was digitized and converted into a database [10] (Fig. 2).

3.1.1 Shoreline Change

The multi-temporal and multi-sensor analysis of satellite images (Landsat-5 and Sentinel-2B) was based on several remote sensing approaches. The automatic detection of shoreline features was done through supervised classification by an object-oriented approach, OBIA, combined with either the Random Forest or Decision Tree Forest RF_Object machine learning algorithms. The automatic extraction method was done in four steps:

Segmentation This allows the capture of well-defined regions or features of an object that can be distinguished from each other. The Mean Shift Segmentation (MSS) algorithm was applied using the Orfeo ToolBox OTB software. Three parameters were determined: spatial radius (sr), range radius (rr), and minimum

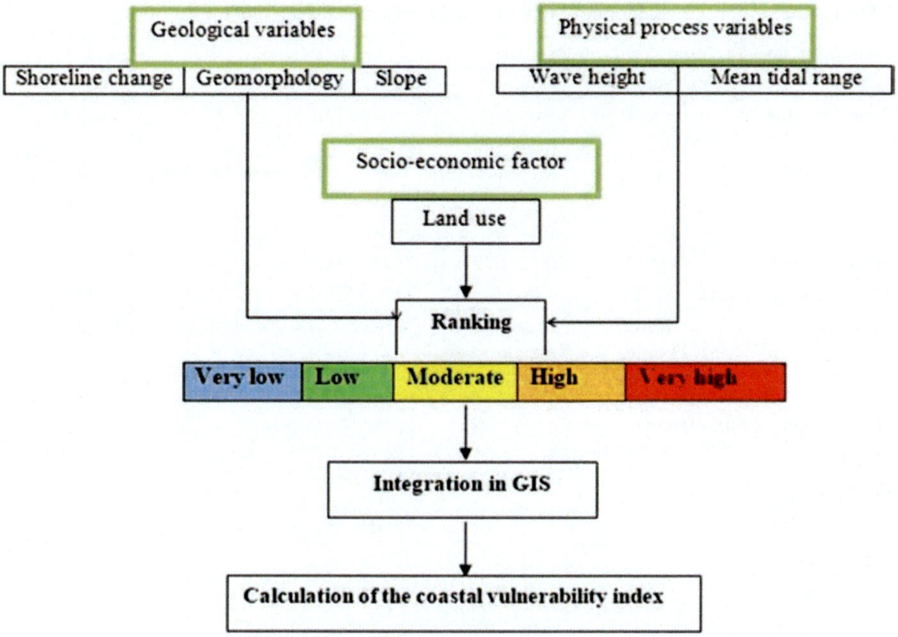

Fig. 2 Working methodology and data processing approach (Table 1). Ranges for the vulnerability ranking of the variables used along the coast of Jerba

region size (mrs). For the Sentinel-2021 image, these parameters were set to 10, 0.05, and 20, respectively. For the Landsat-1989 image, these were set to 10, 0.01, and 5 (Table 2).

Zonal statistics The zonal statistical algorithm of the QGIS software allows the calculation of classification variables for all the parcels contained within a vector layer (QGIS_zonal statistics). Each variable corresponds to the average of the pixel values for each feature. In our case, four statistics were computed: minimum, maximum, average, and standard deviation.

Training and Classification Training was performed using the *TrainVectorClassifier* module of the OTB software. This module enables the Random Forest algorithm to do the classification. Then, the prediction of the classes on the objects resulting from the segmentation was carried out using the Vector Classifier tool.

Cross-validation was performed on the validation samples, which correspond to 20% of the plots in the ground-truth database. A confusion matrix was computed each time to generate the kappa index, which gives global measures of quality, and the global accuracy index, which reflects the classification performance. The extraction of the shorelines themselves was done using a Geographic Information System (GIS). These shorelines have zigzag-like patterns, giving a non-real representation of their geometric nature. The smoothing was done by the Smooth Line tool of the

Table 1 Ranges for the vulnerability ranking of the variables used along the coast of Jerba

Ranking of coastal vulnerability index					
Variables	Very low	Low	Moderate	High	Very high
Shoreline change (m/yr)	>2.0	1.0–2.0	1.0– +1.0	−1.1−−2.0	<−2.0
Geomorphology	Rocky	Medium cliffs	Low cliffs	Cobble beaches	Sand beaches
Slope (%)	>12	12–9	9–6	6–3	<3
Wave height (m)	<0.55	0.55–0.85	0.85–1.05	1.05–1.25	>1.25
Mean tidal range (m)	>6.0	4.0–6.0	2.0–3.99	1.0–1.99	<0.99
Land use	Water bodies	Natural meadow	Forest	Agricultural activity	Urban area

Table 2 Setting parameters

	MSS		
Satellite images	sr	rr	mrs
Landsat 1989	10	0.01	5
Sentinel 2021	10	0.05	20

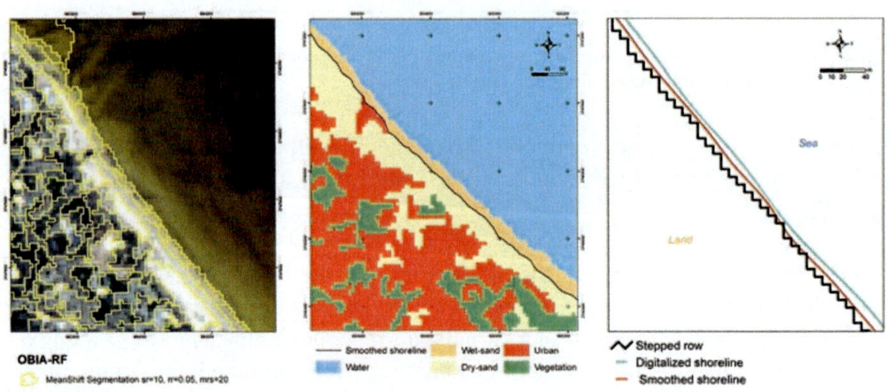

Fig. 3 Shoreline extraction from Sentinel image 15/01/2021

ArcGIS software. Then, the evolution rate was calculated by the End Point Rate (EPR) method of the Digital Shoreline Analysis System extension (DSAS), which allows the measurement of the distance between two successive shorelines. To validate the results, they were compared to those produced (Fig. 3).

3.1.2 Geomorphology

The interaction between hydrodynamic processes and geological formations represents the coastal zone morphology [11]. Therefore, determination of the landforms is very useful in measuring the vulnerability of the studied area. The topographic coverage of the island of Djerba was obtained by digitizing and assembling two digital elevation models (DEMs) obtained from the *Shuttle Radar Topographic Mission* (SRTM) data of the *United States Geological Survey* (USGS). These two DEMs were merged with data from topographic maps at a scale of 1:50,000, covering the coast studied, and were interpolated and correlated under a Geographic Information System (GIS) to develop an altitude map. In addition, field surveys were conducted along the entire coastline to collect data on geomorphological features.

3.1.3 Slope

The slope map was made using the "*Slope*" extension of the ArcGis® 10.4 software from the DEMs. It highlights terrains, whose slope varies between 0° and 2.29°. The slope values were reclassified into four to be representative of the mostly flat terrain in the study area. The majority of the terrain is characterized by a low slope value between 0 and 4%, wherein the average slope value is equal to 3.49%. The low topography extends into the submarine domain by an extensive and shallow continental shelf with an isobath of −10 m located at an average of 3 km from the shore.

3.1.4 Significant Wave Height

Wave height is related to the wind regime. In this study, the data were derived from measurements made by the synoptic station of the airport of Djerba–Zarzis. The annual distribution of winds indicates a high average of east winds (47%), followed by west winds (28%). The eastern coastline of Jerba is dominated by a regime of swells generated by local winds, which means the swell activity is seasonal. In particular, winter swells are quite notable, with the highest coming from the north and the longest swells arriving from the north-northeast and northeast. Waves of more than 4 m high have been recorded in the northeast sector of the island [21]. According to HIDROTECNICA PORTUGUESA (HP 1959), a wave of 4 m, 8 s generates a current of 4 m/s by 2 m depth and 2 m/s by 6 m depth [11].

3.1.5 Mean Tidal Range

Based on the tidal regime, microtidal coasts (<1 m) have a very high vulnerability, and macrotidal coasts (>6 m) a very low vulnerability [28]. The tide all along the coast of Jerba is a regular, semi-diurnal type.

3.1.6 Land Cover

Object-based image analysis OBIA, applied on the Sentinel-2 image (2021), is the method that was adopted to classify our study area. OBIA is based on image objects that are made up of pixels with the same or similar information [7]. Previous field knowledge is necessary in order to perform supervised classification of high-resolution images [26]. In this regard, several field surveys were conducted over different periods during 2018, which gave us the opportunity to visualize the land cover along the coastline. Then, the selection of samples is assessed each time based on a high-resolution Google Earth image. Using this method, regions of interest (ROI) were defined in the Sentinel-2 image: water, wet sand, dry sand, urban, and vegetation. The Mean Shift Segmentation (MSS) algorithm of the Orfeo ToolBox (OTB) software was used to separate the image into segments with high correlation. Finally, the classification was performed by applying the Random Forest (RF) algorithm. A prior optimization of the parameters is essential for each step. The parameters were fixed by the cross-validation method. Two indices were calculated each time: the kappa index, which gives global measures of quality, and the overall accuracy (OA) index, which reflects the classification performance. They respected the order of 0.94 and 0.93, which means that 94% of ground-truth variables were well classified.

3.2 Coastal Vulnerability Index (CVI) Calculation

In the present study, three geological variables (shoreline change, geomorphology, and slope), two physical variables (mean tidal range, significant wave height), and a socio-economic factor (land cover) were used to calculate the coastal vulnerability index. These key variables were then aggregated into a single index, using the following mathematical formula (see Eq. 1). The CVI was applied to quantify the contribution of each variable to the coastal vulnerability of the study area. The CVI is based on the definition of semi-quantitative scores with a scale of 1 to 5 where a value of 1 indicates a low contribution to the coastal vulnerability of a specific key variable for the study area, while a value of 5 indicates a high contribution.

$$CVI = \sqrt{\frac{a.b.c.d.e.f}{6}} \quad (1)$$

where (*a*) is the shoreline change, (*b*) is geomorphology, (*c*) is the slope, (*d*) is the wave height, (*e*) is the mean tidal range, and (*f*) is the land use.

4 Results

The northeastern shoreline, which extends for about 22 km and is highly urbanized, is the most dynamic sector of the island and the most affected by the degradation of the natural environment. These coastal areas are particularly vulnerable due to natural dynamism and extreme storms. The results show that 63% of the shoreline of the northeast and southeast coasts of the island shows a high to a very high degree of vulnerability. It also reveals that the wide sandy coasts (a length of 8 km and width of 700 m) are less vulnerable to erosion risk.

Integrated approaches combining remote sensing and GIS methods have significantly optimized time and effort and provided technological solutions to create a database on coastal areas and their dynamic processes [2].

4.1 The Shoreline Variable

Our study area was divided into two coastal areas to simplify the estimation of shoreline evolution and to facilitate their interpretation, analysis, and presentation; the sandy spit of Ras Rmel and the sandy beaches of the northeast-southeast coast of the island (sector A and sector B) (Fig. 4). The results showed a strong correlation between

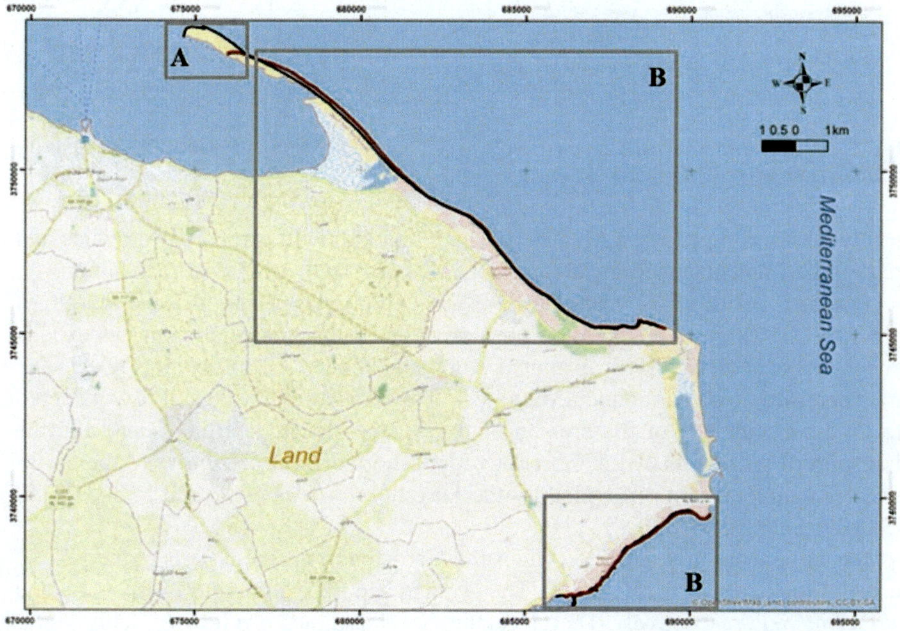

Fig. 4 The eastern coasts of Jerba Island (**a**), Ras Rmel sandy spit, and (**b**) sandy beaches

Table 3 Erosion/accretion rates between 1989 and 2021

	RF_Object		Digitalization	
EPR (1989–2021)	Sandy beaches	Sandy spit	Sandy beaches	Sandy spit
MIN	−4.81	+34.74	−4.06	+37.33
MAX	+1.12	+39.32	+2.37	+39.97
MEAN	−1.8	+37.11	−1.2	+38.24

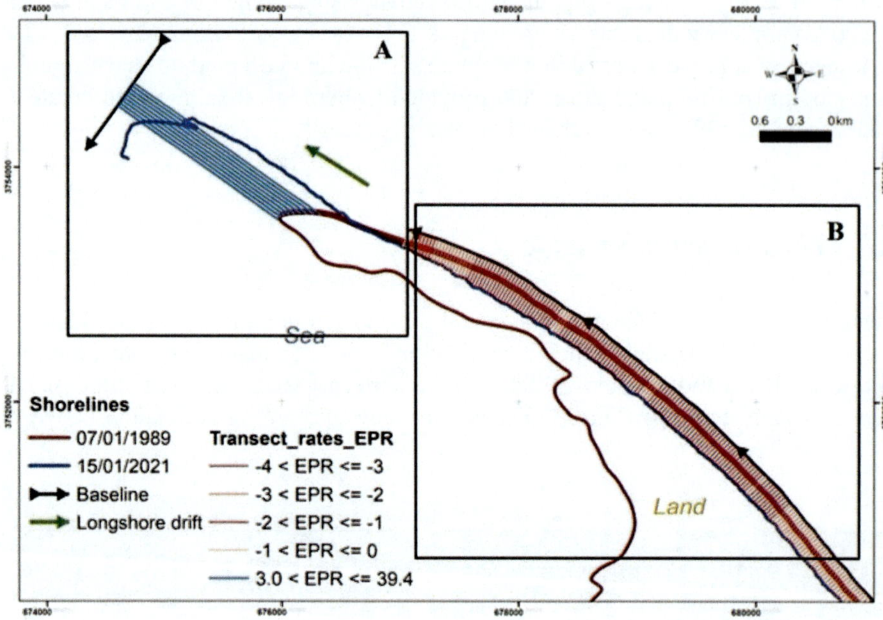

Fig. 5 Results with the OBIA method and calculation of evolution rate with EPR method; (**a**) accretion rate, (**b**) erosion rate

RF_Object and digitization (Table 3), which highlights the strong performance and relevance of machine learning algorithms for coastal land cover class recognition.

During a period of 32 years (1989–2021) and after applying a global margin of error for the adopted methodologies, it is revealed that the sandy spit has experienced strong accretion, with an annual average of about +37.11 m/year (Figs. 5a and 6). Therefore, it was assigned a value of 1. The elongation of the spit is considered to be a consequence of the erosion of the beaches located further south and the transport of sediments by the most active longshore drift directed toward the northwest. The annual input provided by the longshore drift is estimated at 30,000 m^3 (STUDI-SOGREAH, 2002) [11].

Sandy beaches have suffered widespread erosion in different parts of the NE-SE coast of the island (Fig. 5b and 7). The average and maximum rates were − 1.8 and + 1.12 m/year, respectively. Therefore, they were assigned a value of 5. This degradation is explained by the strong anthropic pressure weakening the shoreline. The

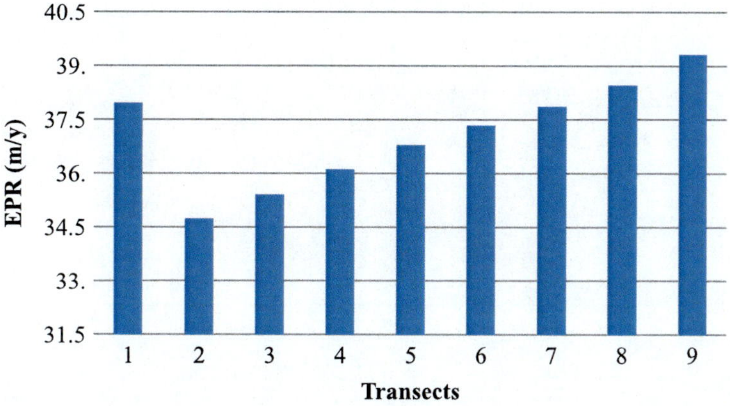

Fig. 6 Evolution rate of Ras Rmel sandy spit between 1989 and 2021

concentration of facilities and seaside activities on the coast makes it vulnerable to an accelerated rise in sea level.

4.2 The Geomorphology Variable

Topographically, Jerba is characterized by a succession of plateaus and plains with an average altitude of 20 m (Fig. 6), with the highest point being Dhahret Guellala, whose southern part reaches 53 m. The plateaus have an irregular surface with a series of slope breaks.

The topography of the plains is defined by a flat and narrow coastal fringe and by low altitudes, often less than 5 m or even 2 m, which explains the existence of sebkhas and chotts (APAL, 2018). Low elevation coastal landforms (<10 m), especially sandy beaches, are assigned a value of 5 (very high vulnerability) [1, 18] (Fig. 8).

4.3 The Slope Variable

Areas with slopes less than 1.5%, including beaches, are more susceptible to sediment transport by longshore drift [1, 12]. The consequences of sea action on a steeply sloping shoreline are negligible, unlike a gently sloping shoreline, where any rise in sea level would flood large areas of land. Therefore, the shoreline is expected to retreat more rapidly on low slopes than on high slopes [1], which is why weighted values of 5 are assigned to our study area (Fig. 9).

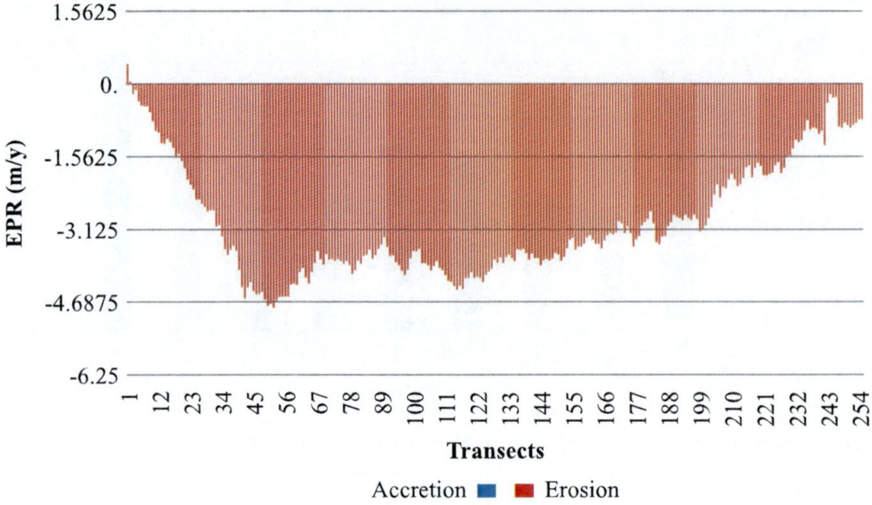

Fig. 7 Evolution rate of the sandy beaches of the northeast and southeast coast of Jerba island between 1989 and 2021

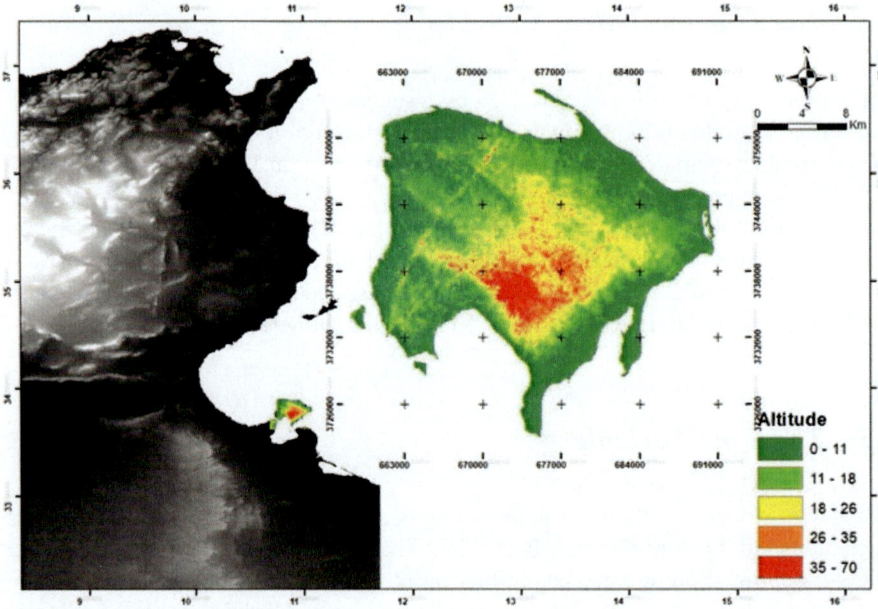

Fig. 8 Altitude map of Jerba

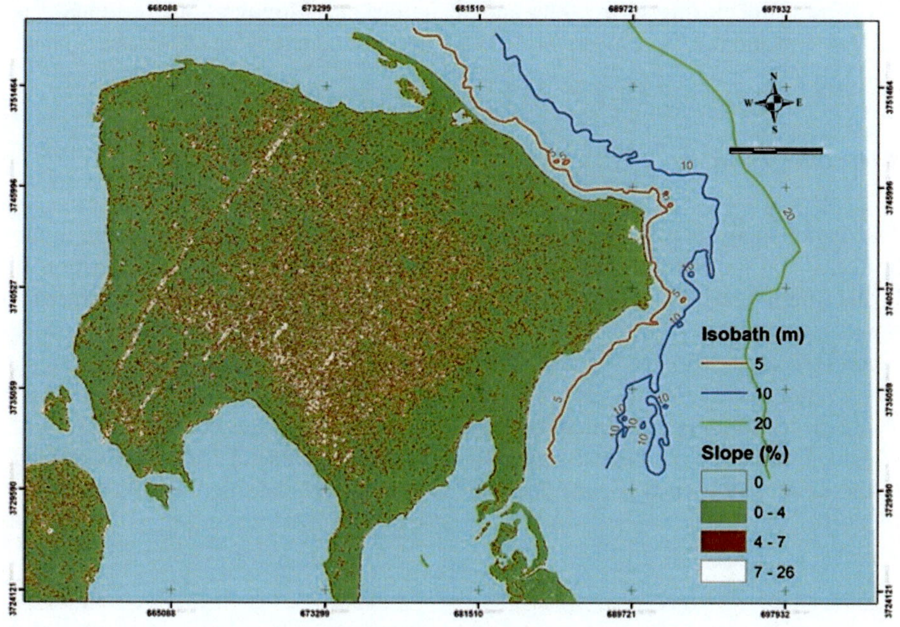

Fig. 9 Map of slopes and bathymetry of the island of Jerba

4.4 The Wave Height Variable

According to general studies of coastal protection against marine erosion (HP, 1995–1997), based on the modeling of winds in the Mediterranean and waves generated on the entire Tunisian coastline, the significant decadal height of swell is 4.8 m over a period of 9.9 s. More recently (IHE, 2007), using the "UK Met Office EuropeanWave Model (UKMO-EWM)" numerical model, which is managed by the British Meteorological Office (MetOffice), the 10-year significant swell height is 4.3 m over a period of 9.5 s. This explains the erosive action of the swell and the longshore drift. Therefore, since coasts with very high significant wave heights are assigned the status of very high risk coasts [25], a value of 5 was assigned in our study (very high vulnerability).

4.5 Mean Tidal Range

The Mediterranean basin exhibits a microtidal regime. According to HP 1995–1997, the southeast sector of the island (Aghir) is characterized by a significant average tide, equal to 0.8 m during spring tides and 0.2 m during neap tides. There is a significant difference in tidal height between the northeast of Djerba and the southeast (Aghir). According to the Environmental Impact Assessment (EIA, 2015) report, the average tide during spring tides and neap tides near Houmet Souk (northeast) is about 1.1 m and 0.2 m, respectively. Therefore, a weighted value of 5 was assigned.

4.6 The Land Use Variable

The overall analysis has shown that the studied shoreline is highly urbanized in the northeast and southeast sectors of the island. This is evident by the multiplication of hotels (more than 130 hotels) and the destruction of the coastal dunes for construction purposes. The shoreline on the southeast side of the island displays a very segmented layout because of the multiplication of groins (more than 25 groins), which, in general, only displaces the damage caused to other neighboring sectors. These developments have increased the vulnerability of the coast. Therefore, a weighted value of 5 has been assigned to them, while a value of 1 has been assigned to the sandy spit, which is considered as a wetland and classified as a Ramsar site (Fig. 10).

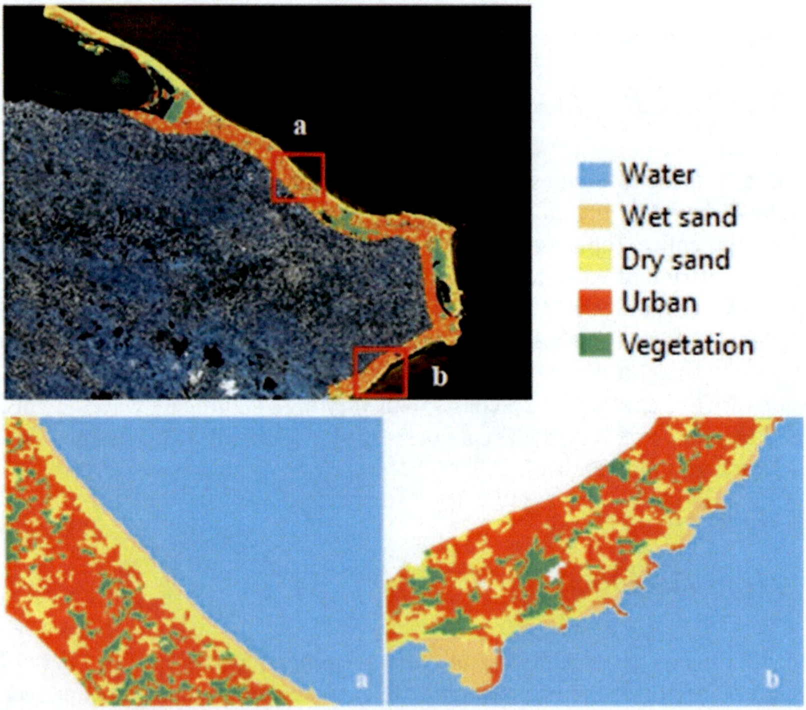

Fig. 10 Northeast (**a**) and southeast (**b**) coastal fringe land use map

4.7 Coastal Vulnerability Index (CVI)

The variables described above each contribute equally to coastal vulnerability. Equation (1), developed by [19], provides a wide range of values. It identifies the variables that are most critical to coastal vulnerability. The lowest and highest values of the CVI are 10.2 and 51.03, respectively (Fig. 11). Following the range generated by this index, our study area, which extends over a linear distance of 22 km, is categorized into four classes: very high, high, moderate, and low vulnerability (Fig. 12).

The results of the statistical classification of the CVI reveal that the very high vulnerability class is the most significant, comprising 45.45% of the coastline (CVI of 51.03). Its heavily urbanized coasts are mainly located in the northeast of the island (Fig. 11a). The coastal areas of high vulnerability correspond to 18.18% of the coastline, concentrated on the sandy spit (CVI of 42.36).

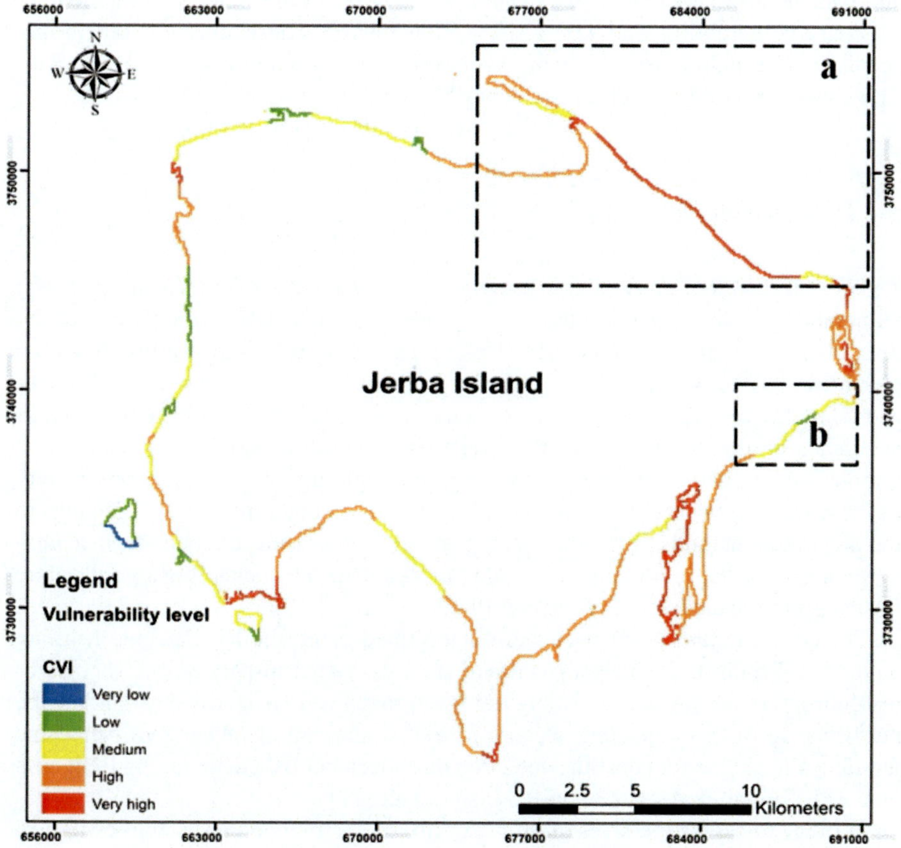

Fig. 11 Coastal vulnerability index of Jerba Island: (**a**) northeast and (**b**) southeast coast

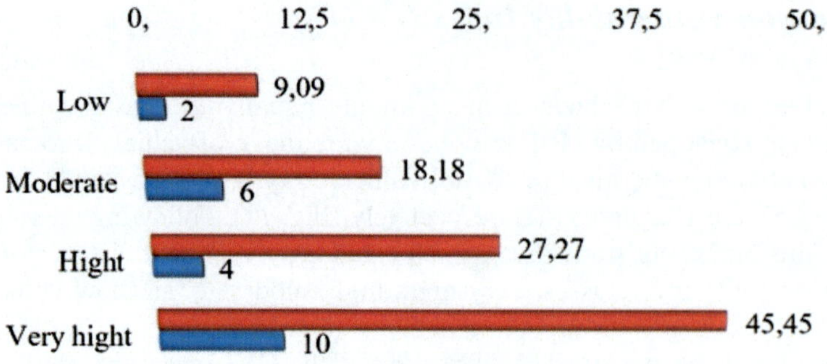

Fig. 12 Graphic presentation of the degree of vulnerability as a percentage of the coastline

The summary map of the CVI reveals that the moderate and low vulnerability classes are mainly in the southeast sector of the island, with values of CVI of 25.9 and 10.2, respectively (Fig. 11b). In addition, 27.27% of the values show a medium degree of vulnerability, and 9.09% show a low degree. These areas are undergoing accretion, the main reason for which could be the continued installation of dikes. There were about twenty dikes for a coastline of about 5 km in 1999 [5].

5 Discussion

The objective of this study was to develop an efficient method for measuring coastal vulnerability and to identify the most vulnerable sites. DVI, complemented by direct fieldwork, appears to be effective in quantifying this vulnerability. It should be noted that the DVI is influenced by both geophysical variables and socio-economic factors. These variables were chosen according to the location and characteristics of the coast [15]. In this work, six variables were processed through remote sensing data. Following a five-degree weighting system, the heterogeneity of these variables was eliminated according to well-defined intervals [1]. Thanks to the integrated methods of remote sensing and GIS, the results obtained from a database on the coastal relief and its dynamic mechanisms [13], enabling a good future for integrated management of the coastline.

The coastline and land cover were determined using the RF machine learning algorithm. Preliminary analyses via field surveys, visual inspections of very-high-resolution satellite images, and a digitizing approach were performed to validate the performance of this algorithm. Several recent studies on the accuracy of machine learning algorithms for coastline detection have been published, including Bengoufa et al. [2], Kumar et al. [16], and McAllister et al. [17].

In order to assess the level of vulnerability of socio-economic activities to erosion risk and to obtain more in-depth results, the land use variable was incorporated into the CVI. This allowed us to promote good coastal planning in order to foster sustainable coastal management and development.

Nevertheless, this study highlights the lack of data, especially very high resolution data and tide gauge data, and the need for further research. These limitations are compounded by the fact that the results of the index calculation do not offer the accuracy of detailed field studies. Nevertheless, fieldwork, remote sensing, and GIS allow for the development of a local CVI tailored to field conditions.

Vulnerability refers to the ability of territories to respond to hazards, as well as their ability to adapt in the longer term [4]. In our case study, and with a perspective of spatialization of risks, we attempt to assess the vulnerability of the coastline of the island of Jerba, facing the risk of coastal erosion.

This assessment shows a very large distribution of vulnerability especially in the northeast sector of the island. This sector has undergone the strongest urbanization since the tourist boom experienced by the region. The developments on the seafront are massive and continuous [23]. The destruction of coastal dunes, as well as beach and tourist activities, has stopped the transport of continental sand that naturally nourishes the beaches. Therefore, the damage and consequences of coastal erosion will be all the more significant. The sandy beaches, in the absence of preventive measures, will therefore be exposed to increasingly intense erosion leading to their disappearance. If coastal preservation in this region is the goal, it will be necessary in the future to define a distance to be respected for the hard constructions on the shore. This will affect future approaches in management and the redevelopment of the public maritime domain.

Our study area suffers from numerous hazards, a high concentration of impact factors on the coastal fringe, weak risk management in the field of urban planning, and the lack of awareness of these issues of the inhabitants. The CVI remains a basic index that can be improved by integrating other types of variables such as structural factors that include risk management and the relationship of the population to risk.

6 Conclusion

The integrated methods of remote sensing and GIS and the supervised classification by an object-oriented approach combined with the Random Forest machine learning algorithm (OBIA-RF) have provided basic data on the physical processes of coastal areas and land use.

The coastal vulnerability index adapted to the Mediterranean coasts, which takes into account six variables, shows its efficacy in obtaining a quantitative measure of coastal vulnerability.

Based on geophysical and socio-economic variables processed by remote sensing methods, this work presents the results of an assessment of the vulnerability repartition of the NE-SE coast of Jerba and highlights the areas that are likely to be the most affected by future sea-level rise. Four categories of vulnerability, extracted from the sum of variables, have been classified: low, medium, high, and very high vulnerability.

We have found that 63% of the coastline of the northeast coast of the island has a high to very high degree of vulnerability. Moreover, 37% of the coastline of the southeast of the island has a low to moderate vulnerability.

The CVI allows scientists to assess the possibility of physical changes due to continued sea-level rise along the coastal areas. In particular, it is of great importance in that it can provide a good database for prospective studies of sustainable management projects. These results can, however, be improved for future purposes by feeding other types of data into the CVI calculation sub-indices, in combination with very high spatial resolution satellite data. Indeed, the assessment of the vulnerability degree enables an understanding of the characteristics of coastal environments for a good integrated management.

Acknowledgments The authors thanks the ERASMUS+ program and the Jean Monnet Chair for the European Spatial Studies of Sea and Coastal zones -599967-EPP-1-2018-1-FR-EPPJMO-CHAIR.

References

1. Bengoufa S, Niculescu S, Mihoubi K, Belkessa R, Abbad K (2021) Cartographie de la vulnérabilité côtière par l'intégration des méthodes géo-spatiales: cas de Mostaganem (ouest Algérien):14
2. Bengoufa S, Niculescu S, Mihoubi MK, Belkessa R, Rami A, Rabehi W, Abbad K (2021) Machine learning and shoreline monitoring using optical satellite images: case study of the Mostaganem shoreline. Algeria J Appl Rem Sens 15. https://doi.org/10.1117/1.JRS.15.026509
3. Bera R, Maiti R (2019) Quantitative analysis of erosion and accretion (1975–2017) using DSAS — a study on Indian Sundarbans. Reg Stud Mar Sci 28:100583. https://doi.org/10.1016/j.rsma.2019.100583
4. Birkmann J, Welle T (2015) Assessing the risk of loss and damage: exposure, vulnerability and risk to climate-related hazards for different country classifications. IJGW 8:191. https://doi.org/10.1504/IJGW.2015.071963
5. Bourgou M (2005) Les plages: impact des aménagements touristiques et portuaires sur leur évolution récente: exemples tunisiens. République tunisienne, Ministère de la recherche scientifique et de la technologie et du développement de compétences, Centre d'etudes et de recherches economiques et sociales
6. Chen C, Bu J, Zhang Y, Zhuang Y, Chu Y, Hu J, Guo B (2019) The application of the tasseled cap transformation and feature knowledge for the extraction of coastline information from remote sensing images. Adv Space Res 64:1780–1791. https://doi.org/10.1016/j.asr.2019.07.032
7. Eriksen SH, Kelly PM (2007) Developing credible vulnerability indicators for climate adaptation policy assessment. Mitig Adapt Strat Glob Change 12:495–524. https://doi.org/10.1007/s11027-006-3460-6
8. Gargouri D, Gzam M, Kharroubi A, Jedoui Y (2018) Use of sediment quality indicators for heavy metals contamination and ecological risk assessment in urbanized coastal zones. Environ Earth Sci 77:381. https://doi.org/10.1007/s12665-018-7567-3
9. Gornitz V, White TW, Cusman RM (1991) Vulnerability of the US to future sea level rise, Coastal Zone, Proceedings of the 7th Symposium on Coastal and Ocean Management, American Society of Civil Engineers, pp 1345–1359

10. Gzam M, Mansouri B, Gargouri D, Kharroubi A (2021) Assessment of the coastal sensitivity using the CS index in the Gulf of Gabes Tunisia; Southern Mediterranean Coast (preprint). In Review. https://doi.org/10.21203/rs.3.rs-1098137/v1
11. Gzam M, Souissi R, Choura M, Jedoui Y (2009) Caractérisation morphologique et sédimentologique de la côte orientale de l'île de Jerba (Sud-Est tunisien). In: Edition 1, Hammamet, Tunisie. Presented at the Conférence Méditerranéenne Côtière et Maritime – Coastal and Maritime Mediterranean Conference, Editions Paralia, pp 99–102. https://doi.org/10.5150/cmcm.2009.025-2
12. Jeyabal G, Sooriamuthu R (2014) Down core variation in sediment characteristics and trace element geochemistry of a core sample in Pichavaram Mangrove area, Tamil Nadu, Southeast Coast of India
13. Kaliraj S, Chandrasekar N, Magesh NS (2015) Evaluation of coastal erosion and accretion processes along the southwest coast of Kanyakumari, Tamil Nadu using geospatial techniques. Arab J Geosci 8:239–253. https://doi.org/10.1007/s12517-013-1216-7
14. Konko Y, Okhimambe A, Nimon P, Asaana J, Rudant JP, Kokou K (2020) Coastline change modelling induced by climate change using geospatial techniques in Togo (West Africa). ARS 09:85–100. https://doi.org/10.4236/ars.2020.92005
15. Koroglu A, Ranasinghe R, Jiménez JA, Dastgheib A (2019) Comparison of coastal vulnerability index applications for Barcelona Province. Ocean & Coastal Manage 178:104799. https://doi.org/10.1016/j.ocecoaman.2019.05.001
16. Kumar L, Afzal MS, Afzal MM (2020) Mapping shoreline change using machine learning: a case study from the eastern Indian coast. Acta Geophys 68:1127–1143. https://doi.org/10.1007/s11600-020-00454-9
17. McAllister E, Payo A, Novellino A, Dolphin T, Medina-Lopez E (2022) Multispectral satellite imagery and machine learning for the extraction of shoreline indicators. Coast Eng 174:104102. https://doi.org/10.1016/j.coastaleng.2022.104102
18. Mclaughlin S, Cooper JAG (2010) A multi-scale coastal vulnerability index: a tool for coastal managers? Environ Hazards 9:233–248. https://doi.org/10.3763/ehaz.2010.0052
19. McLaughlin S, McKenna J, Cooper JAG (2002) Socio-economic data in coastal vulnerability indices: constraints and opportunities. J Coast Res 36:487–497. https://doi.org/10.2112/1551-5036-36.sp1.487
20. Mihoubi MK, Belkessa R, National School for Marine Sciences and Coastal Management (ENSSMAL). Algeria, Latreche, M.A., Laboratory Mobilization and Recovery of Water Resource (LMVR-ENSH), National School for Hydraulics (ENSH), Algeria (2014) Study of the Vulnerability of Coastal Areas of the Algerian Basin with the GIS. IJESD 5:522–526. https://doi.org/10.7763/IJESD.2014.V5.538
21. Miossec J-M, Paskoff R (1979) Evolution des plages et aménagements touristiques à Jerba (Tunisie): Le cas du littoral nord-est de l'île. medit 35:99–106. https://doi.org/10.3406/medit.1979.1903
22. Oueslati A (2004) Littoral et aménagement en Tunisie: des enseignements de l'expérience du vingtième siècle et de l'approche géoarchéologique à l'enquête prospective. Publications de la Faculté des Sciences Humaines et Sociales, Tunis
23. Oueslati A, Labidi O, Elamri T (2015) Atlas de la vulnérabilité du littoral tunisien à l'élévation du niveau marin
24. Pirazzoli PA (1986) Secular trends of relative sea-level (RSL) changes indicated by tide-gauge records. J Coast Res:1–26
25. Pramanik MK, Dash P, Behal D (2021) Improving outcomes for socioeconomic variables with coastal vulnerability index under significant sea-level rise: an approach from Mumbai coasts. Environ Dev Sustain 23:13819–13853. https://doi.org/10.1007/s10668-021-01239-w
26. Sangare H (2020) Évolution de l'occupation du sol dans le bassin versant de Korola (région de Sikasso, Mali) à partir des images satellitaires landsat 15
27. Statistiques | INS [WWW Document] (n.d.) URL http://www.ins.tn/statistiques/111. Accessed 2.23.22

28. Thieler RE, Hammar-Klose ES (2000) National assessment of coastal vulnerability to sea-level rise: preliminary results for the U.S. Atlantic Coast, U.S. open-file report 99-593, Title Page [WWW ocument]. http://pubs.usgs.gov/of/2000/of00-179/. https://pubs.usgs.gov/of/1999/of99-593/. Accessed 3.10.22
29. Masson-Delmotte V, Zhai P, Pirani A, Connors SL, Péan C, Chen Y, Goldfarb L, Gomis MI, Robin Matthews JB, Berger S, Huang M, Yelekçi O, Rong Y, Zhou B, Lonnoy E, Maycock TK, Waterfield T, Leitzell K, Caud N (2021) Climate change 2021 the physical science basis. Front Cover Artwork, Switzerland
30. Wang Y, Liu Y, Jin S, Sun C, Wei X (2019) Evolution of the topography of tidal flats and sandbanks along the Jiangsu coast from 1973 to 2016 observed from satellites. ISPRS J Photogramm Remote Sens 150:27–43. https://doi.org/10.1016/j.isprsjprs.2019.02.001
31. Yan D, Yao X, Li J, Qi L, Luan Z (2021) Shoreline change detection and forecast along the Yancheng coast using a digital shoreline analysis system. Wetlands 41:47. https://doi.org/10.1007/s13157-021-01444-3

A Random Forest Approach for Evaluating Forest Cover Changes Outside Dikes with Sentinel Images

Nguyen Chi Lam, Hiep Xuan Huynh, Simona Niculescu, Quynh Do Nguyen, and Ngan Chau Vo Nguyen

Abstract The dike surrounding the coast of the Vietnamese Mekong Delta (VMD) helps prevent saltwater intrusion and coastal erosion. In order to preserve these dikes, an area of protective forest is planted outside them, helping to maintain and strengthen them. Therefore, monitoring the change in the area of planted forest outside the dike will help to assess the stability of and possible threats to the coastal area and ecosystem. In this paper, using Sentinel remote sensing images, we propose a new approach; applying the Random Forest technique to assess the changes in the planted forest outside the dike. The experimental results obtained on a typical coastal area of the Vietnamese Mekong Delta will help to clearly identify the threats and the evolution of the coastline through the changes in forest area outside the dike.

Keywords Forest cover · Random forests · Sea dike · Sentinel image · Thread

1 Introduction

The Mekong Delta plays an important role in both agriculture and aquaculture for the whole of Vietnam. However, the coastal zone of the Vietnamese Mekong Delta (VMD) region is sensitive to the impact of not only human activities but also climate change [6]. These impacts are predicted to cause increased intensity and

N. C. Lam · H. X. Huynh · Q. Do Nguyen · N. C. V. Nguyen (✉)
Can Tho University, Can Tho City, Vietnam
e-mail: lcnguyen@ctu.edu.vn; hxhiep@ctu.edu.vn; ndquynh@ctu.edu.vn; nvcngan@ctu.edu.vn

S. Niculescu
Laboratory LETG-Brest, UMR 6554 CNRS, University of Western Brittany, Plouzané, France
e-mail: simona.niculescu@univ-brest.fr

S. Niculescu (ed.), *European Spatial Data for Coastal and Marine Remote Sensing*, https://doi.org/10.1007/978-3-031-16213-8_8

frequency of erosion and/or accretion in the coastal zone of the delta [11]. Several researchers have carried out studies on the erosion and/or accretion processes and subsequently suggested the solutions required to minimize any damage to infrastructure and farmland caused by sediment dynamics [13]. Besset et al. [1] confirmed that the coastal mangrove forest plays a key role in sea dike protection and land preservation. In this context, satellite images were used to assess the transformation and impact of any implementation projects to the coastal zones, in general, and the VMD coastal zone in particular [2, 15].

Monitoring land use changes by using satellite images is an oft-used approach [7, 8]. However, the combined technique of using radar and optic data may be challenging in some land cover studies, especially in forest cover studies. This paper aims to monitor the change in coastal mangrove forest cover through the use of bagging—Random Forest algorithms using Sentinel images—and then determine the optimal method to assess the forest cover changes by time series analysis.

This study focuses on the coastline of the Bac Lieu Province, which is located in the southeast coastal area in the VMD. According to Lappe et al. [9], Bac Lieu Province is the third ranking in coastline erosion among the seven coastal provinces in the delta.

The paper is organized as follows: Section 2 presents the Sentinel-1 and Sentinel-2 images along with the pre-processing techniques performed. Section 3 introduces the methodology used to transform the Sentinel images into the matrix data with the rows representing pixels and the columns representing the bands/attributes obtained. Section 4 models the matrix data obtained from using a machine learning algorithm, Random Forest, on the Sentinel images. Section 5 models some important scenarios to evaluate the forest restoration and deforestation in three recent years. Finally, we conclude and present some discussions for future research.

2 Sentinel Imagery

2.1 Sentinel-1

The Sentinel-1 mission currently consists of two polar-orbiting satellites, Sentinel-1A and Sentinel-1B (1B currently inactive), which operate day and night, sensing with a C-band synthetic aperture radar instrument operating at a center frequency of 5.405 GHz, allowing the acquisition of imagery regardless of weather or light conditions. There are future plans to add to the satellite network, providing data until at least 2037. The Sentinel-1 satellite network acquires synthetic aperture radar (SAR) data in single or dual polarization with a revisit time of 6 days. The two types of Sentinel-1 level-1 data produced are distributed by the Copernicus Open Access Hub: Ground Range Detected (GRD) and Single Look Complex (SLC).

The standard generic workflow to pre-process Sentinel-1 GRD data is presented in Fig. 1. The workflow was created in order to be used within the Sentinel

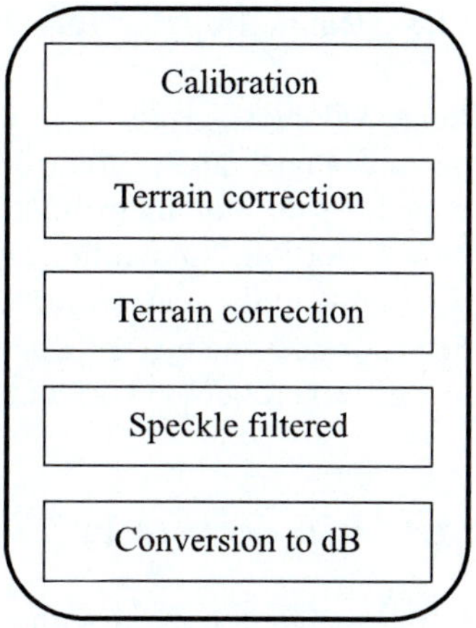

Fig. 1 Pre-processing the Sentinel-1 images with the SNAP toolbox

application platform (SNAP), a common architecture for all Sentinel satellite toolboxes. The pre-processing workflow consists of four processing steps, designed to best reduce error propagation in subsequent processes.

2.1.1 Calibration

Calibration is the procedure that converts digital pixel values to radiometrically calibrated SAR backscatter. The information required to apply the calibration equation is included within the Sentinel-1 GRD product. Specifically, a calibration vector included as an annotation in the product allows for the simple conversion of image intensity values into sigma nought values. The calibration reverses the scaling factor applied during level-1 product generation and applies a constant offset and a range-dependent gain, including the absolute calibration constant.

2.1.2 Speckle Filtering

Speckle, appearing in SAR images as granular noise, is due to the interference of waves reflected from many elementary scatters [10]. Speckle filtering is a procedure wherein the image quality is increased by reducing speckle. When such a procedure is done at an early SAR data processing stage, speckle is not propagated in subsequent processes (i.e., terrain correction or conversion to dB). Speckle filtering is not advisable when there is an interest in the identification of small spatial structures or image texture, since it might remove such information. For visual interpretation, the refined filter has been found to be superior compared to other single product speckle filters because of its ability to preserve edges, linear features, and point target and texture information [10].

2.1.3 Range Doppler Terrain Correction

Range Doppler Terrain Correction SAR data are generally sensed with a varying viewing angle greater than 0 degrees, resulting in images with some distortion related to side-looking geometry. Terrain corrections are intended to compensate for these distortions so that the geometric representation of the image will be as close as possible to the real world. Range Doppler Terrain Correction corrects geometric distortions caused by topography, such as foreshortening and shadows, using a digital elevation model to correct the location of each pixel.

2.1.4 Conversion to dB

As the last step of the pre-processing workflow, the unitless backscatter coefficient is converted to dB using a logarithmic transformation.

2.2 Sentinel-2

Sentinel-2 is a European wide-swath, high-resolution, multispectral imaging mission. The Sentinel-2 array consists of two satellites: Sentinel-2A was launched in 2015 and Sentinel-2B in 2017. Two additional satellites (Sentinel-2C and Sentinel-2D) are slated to launch in 2024. These two additional satellites will cut the revisit time in half. The full mission, with the twin satellites flying in the same orbit but phased at 180°, is designed to give a high revisit frequency of 5 days at the equator. The orbital swath width is 290 km.

The twin satellites of Sentinel-2 provide continuity of Spot and Landsat-type image data, contribute to ongoing multispectral observations, and are used to the benefit of worldwide services and applications such as land management, agriculture and forestry, disaster control, humanitarian relief operations, risk mapping, and security.

2.2.1 Sentinel-2 Bands

The Sentinel-2 array carries the multispectral imager (MSI). This sensor delivers 13 spectral bands ranging from 10 to 60 m pixel size (Table 1).

1. The blue (B2), green (B3), red (B4), and near-infrared (B8) channels have a 10 m resolution.
2. Next, its red edge (B5), near-infrared NIR (B6, B7, and B8A), and shortwave infrared SWIR (B11 and B12) have a ground sampling distance of 20 m.
3. Finally, its coastal aerosol (B1) and cirrus bands (B10) have a 60 m pixel size.

Table 1 Sentinel-2 bands

Band	Resolution (m)	Central wavelength (nm)	Description
B1	60	443	Ultra blue (coastal and aerosol)
B2	10	490	Blue
B3	10	560	Green
B4	10	665	Red
B5	20	705	Visible and near infrared (VNIR)
B6	20	740	VNIR
B7	20	783	VNIR
B8	10	842	VNIR
B8a	20	865	VNIR
B9	60	940	Shortwave infrared (SWIR)
B10	60	1375	SWIR
B11	20	1610	SWIR
B12	20	2190	SWIR

2.2.2 Sentinel Band Combinations

The use of specific band combinations enables the extraction of specific information from an image. For example, there are band combinations that can highlight geological, agricultural, or vegetation features in an image. The Sentinel Playground allows the online testing of different band combinations. Below are described two of the band combinations that were most relevant to our work, though other combinations are certainly possible.

Natural Color (B4, B3, B2)

The natural color band combination uses the red (B4), green (B3), and blue (B2) channels. Its purpose is to display imagery the same way our eyes see the world ("true color"). Healthy vegetation appears green, while urban features often appear white and grey. Finally, water is a shade of dark blue depending on how clean it is (Fig. 2).

Color Infrared (B8, B4, B3)

The color infrared band combination is meant to emphasize healthy and unhealthy vegetation. The near-infrared (B8) band is sensitive to chlorophyll reflection. In the frequently used false-color band combination R, G, B and B8, B4, B3, the NIR band is displayed on the red channel, leading to vegetation being shown in red in this combination, bare soil and urban features in white, and water in blue (Fig. 3).

Fig. 2 Natural color on the coastal zone of Bac Lieu Province on the Sentinel-2 image acquired on January 31, 2022

Fig. 3 Color infrared on the Bac Lieu coastline on January 31, 2022

3 From Sentinel Data to Matrix

The satellite image data needs to be converted into two-dimensional row and column matrix data. Each row is an object to be analyzed (pixel, object), while each column is the value of the features (spectral value, vegetation value, backscatter value, observation data, etc.). Figure 4 describes an algorithm that converts data from remote sensing images into a matrix dataset for image analysis.

In order to improve the analysis results for this paper, the normalized difference vegetation index (NDVI) was used and integrated with the data into the dataset.

$$\mathrm{NDVI} = \frac{\mathrm{NIR} - \mathrm{Red}}{\mathrm{NIR} + \mathrm{Red}} \tag{1}$$

The NDVI value is calculated from the difference between the infrared and red spectrum holding values. That value represents the density of vegetation; the higher the value (closer to 1) the higher the density, and vice versa, the lower the value (closer to −1), the less vegetation there is at the specified location [14].

The matrix is two-dimensional, $m \times n$, where:

Row $P = \{p_1, p_2, p_3 \ldots p_M\}$, p_k is a pixel in images with n value attribute.
Column $A = \{a_1, a_2 \ldots a_n\}$, a_j is an attribute of pixels.

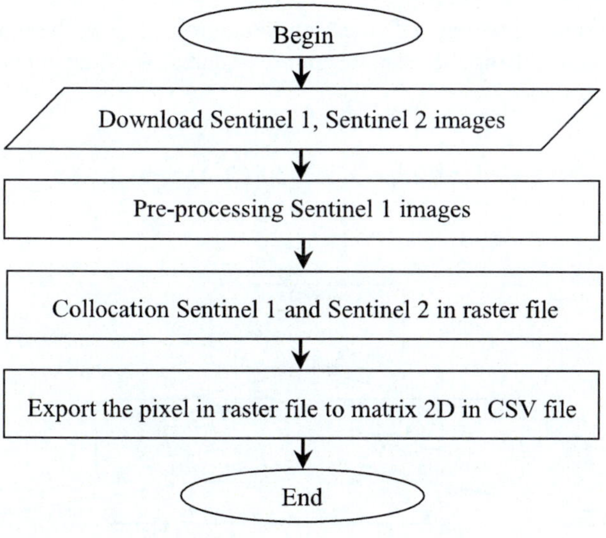

Fig. 4 From Sentinel data to matrix

In this paper, there are nine columns in the matrix:

- Blue S2B2 Band 2 of Sentinel-2
- Green S2B3 Band 3 of Sentinel-2
- Red S2B4 Band 4 of Sentinel-2
- NIR S2B8 Band 8 of Sentinel-2
- Elevation srtm 3Sec SRTM elevation
- Sigma VV S1 VV Polarization VV
- Sigma VH S1 VH Polarization VH
- NDVI Normalized difference Vegetation index
- Class Observation data Others or forest

4 Sentinel as a Random Forest

4.1 Decision Tree

Each pixel is represented as a 9-tuple information packet with a decision attribute (CLASS). The other information is the conditional attributes of ID, BLUE, GREEN, RED, NIR, ELEVATION, VH, VV, and NVDI.

The value of each attribute is what partitions the remote sensing image into different regions [4, 5]. The decision tree/classification tree obtained is used to predict the pixel class that most commonly occurs in the training pixel dataset (region partitioned with the same class) (Fig. 5).

The remote sensing region is divided into predictor space, with distinct and non-overlapping regions, using the residual sum of squares (RSS) error calculation. For example, with Q regions:

$$\mathrm{RSS} = \sum_{q=1}^{Q} \sum_{i \in \mathrm{region}(q)} \left(\mathrm{CLASS}_i - \overline{\mathrm{CLASS}}_{\mathrm{region}(q)} \right)^2 \tag{2}$$

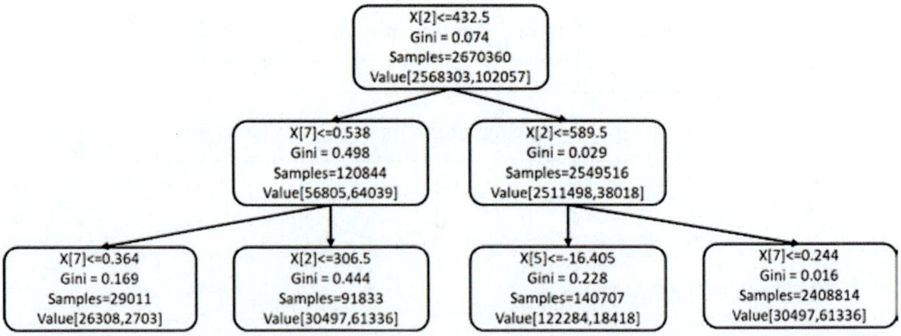

Fig. 5 An example of a decision tree obtained from a remote sensing dataset

The splitting indices for each conditional attribute are the Gini index or entropy.

As $\widehat{p_{qk}}$ is the proportion of training pixels in the q_{th} region of k_{th} class. The Gini value is computed by:

$$\text{Ginivalue} = \sum_{q=1}^{Q} \widehat{p_{qk}} \left(1 - \widehat{p_{qk}}\right) \tag{3}$$

The entropy value is computed by:

$$\text{Entropyvalue} = \sum_{q=1}^{Q} \widehat{p_{qk}} \log\left(\widehat{p_{qk}}\right) \tag{4}$$

4.2 *Bootstrap*

The bootstrap method [5] is used to estimate the standard errors of the coefficient from a linear regression fit. The bootstrap is a sampling technique with replacement (each pixel can occur more than one in the bootstrap dataset).

If B is a remote sensing dataset and T bootstraps are produced, $B^1, B^2, B^3 \dots B^T$ with T corresponding estimates of $\hat{e}^1, \hat{e}^2, \hat{e}^3 \dots \hat{e}^T$, then the standard error of these bootstraps is computed as:

$$\text{Standarderror}_T\left(\hat{e}\right) = \sqrt{\frac{1}{T-1} \sum_{i=1}^{T} \left(\hat{e}^i - \frac{1}{T} \sum_{j=1}^{T} \hat{e}^j\right)^2} \tag{5}$$

4.3 *Bagging*

Bagging, or bootstrap aggregation [3, 5] is used with remote sensing data to reduce the high variance of the decision trees obtained. In other words, we compute the average of a set of pixels in order to reduce variance (high variance will occur if we split the training dataset into parts randomly). Each bootstrap is used to construct a decision tree with all the attributes and a majority vote is performed.

With T bootstraps and $\widehat{f^i}\left(\text{trainingdata}\right)$ being the prediction of a training dataset, the final predictions will be computed as:

$$\overline{\text{f}_{\text{bagging}}\left(\text{trainingdata}\right)} = \frac{1}{T} \sum_{i=1}^{T} \widehat{f^i}\left(\text{trainingdata}\right) \tag{6}$$

4.4 Random Forest

The Random Forest algorithm [3, 5] is used for bagging, with random sample attributes equivalent to the square root of the number of total attributes.

5 Experiment

5.1 Dataset

Bac Lieu's coastal area has a length of 54.6 km, where erosion and sedimentation both occur erratically by natural and human factors. Both the government and the people have proposed projects to protect the local coastal area, including planting projects and the restoration of protective forests outside the sea dike of Bac Lieu Province. Due to the frequent changes in protective coastal forests, we have selected the Hiep Thanh and Vinh Trach Dong Communes of Bac Lieu Province as the main regions of interest, as seen in Fig. 6.

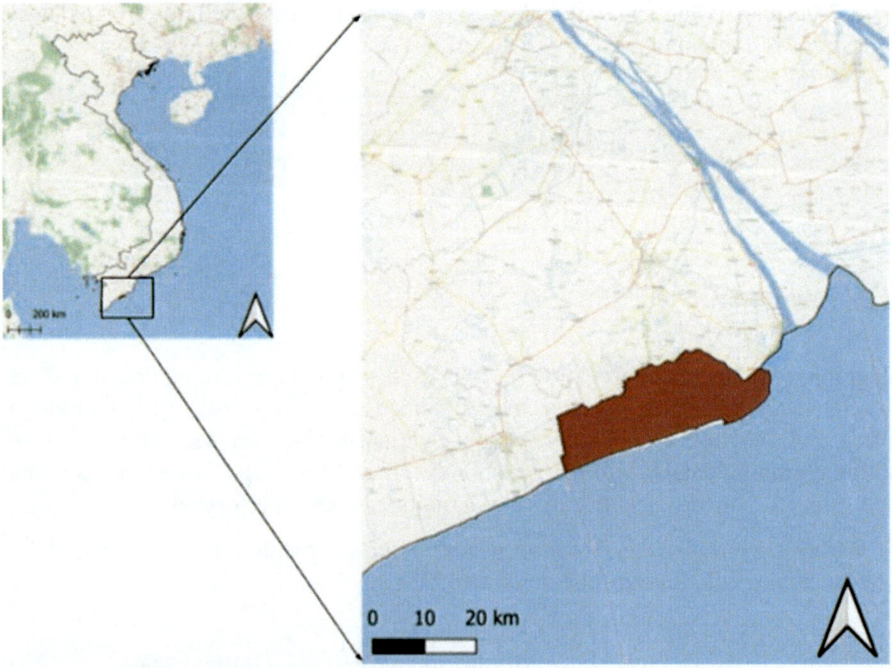

Fig. 6 Coast of Bac Lieu Province

Table 2 Acquisition dates of used Sentinel data

Year	Sentinel-1	Sentinel-2
2019	31/01/2019	31/01/2019
2020	21/01/2020	21/01/2020
2021	08/02/2021	09/02/2021
2022	20/01/2022	21/01/2022

Table 3 An example of the matrix dataset

id	Blue	Green	Red	NIR	Elevation	VV	VH	NDVI	Class
0	588	1040	984	962	−6.637	0	0	−0.011	0
1	617	1080	1180	1003	−6.362	0	0	−0.081	0
2	559	966	1002	659	−6.090	0	0	−0.206	0
3	311	557	502	2414	−1.424	0	0	0.655	1
4	193	436	340	2510	−1.395	0	0	0.761	1
5	303	564	582	2434	−1.735	−16	−10	0.614	1

Cloud-free satellite image data was collected during the dry season in the VMD region. Between 2019 and 2022, one optical image and one radar image were collected (Table 2).

Through pre-processing, optical and radar data were integrated into one dataset per year, leading to four datasets corresponding to the four observation years of 2019 to 2022. Each dataset has nine columns corresponding to three visible color bands, one near infrared band, altitude data, two polarization VV and VH, vegetation index (NDVI), and the observation column (class) as presented in Table 3.

5.2 *Tools*

The SAR images were pre-processed to different formats using the SnAPS toolbox. QGIS is used to construct ground-truth images for the process of generating monitoring data and displaying results. The Random Forest algorithm was run in Python.

6 Results

6.1 *Accuracy of the Random Forest Model by Year*

A machine learning model, bagging Random Forest, was built in Python and trained with the datasets of each year from 2019 to 2022. The accuracy of the model is shown in Table 4. Compared to previous studies, the accuracy in this study is very high and in line with related studies.

Table 4 Accuracy of the Random Forest model from 2019 to 2022

Year	Other areas (%)	Forest area (%)	Overall (%)
2019	99.63	64.84	98.17
2020	99.61	66.88	98.28
2021	99.66	71.46	98.64
2022	99.68	65.53	98.53

Table 5 Estimated forest area by year of 2019 to 2022

Year	Forest area (ha)	Others areas (ha)
2019	402.59	9134.41
2020	388.60	9148.40
2021	347.45	9189.55
2022	321.87	9215.13

6.2 Forest Area by Year from 2019 to 2022

According to Table 5, the area of protective forest has decreased year by year. This corresponds with the fact that protective forest is very difficult to preserve, although the government has implemented many forest restoration projects. The protective forest in this area is strongly affected by the wave impact from the sea.

Comparing pairs of images from two consecutive years, the types and rates of changes were assessed. The observed changes were restored forest areas, lost forest areas, and the areas that remained constant. The highest forest loss occurred between 2020 and 2021, with 10.6%, while the lowest forest loss rate of 3.5% occurred between 2020 and 2019.

6.3 Forest Restoration from 2020 to 2022

The area of restored forest between 2020 and 2022 is shown in Table 6. From the results, we can see that the restored forest area gradually decreases through the years 2019–2021 but then recovers strongly in 2021–2022. However, based on the ratio between the restored forest area and the total forest area of each year, the ratios slightly increase from year 2020 to 2021 with more than 3% and then jump up two-fold in 2022.

Indeed, on April 1, 2021, the Vietnamese prime minister issued Decision No. 524/QD-TTg, approving the "Planting a billion trees from 2021 to 2025" project [16]. According to the project, by the end of 2025, at least one billion trees will be planted, of which 690 million trees will be scattered in urban and rural areas, and 310 million trees concentrated in protective forests, specialized forests, and new production forests. The aim of the project is to contribute to the protection of the

Table 6 The forest area restored from 2020 to 2022

Year	Restoration area (ha)	Forest area (ha)	Ratio (%)
2020	11.90	388.60	3.06
2021	10.83	347.45	3.12
2022	20.82	321.87	6.47

Table 7 The deforestation from 2020 to 2022

Year	Deforestation area (ha)	Forest area (ha)	Ratio (%)
2020	25.89	388.60	6.66
2021	51.98	347.45	14.96
2022	46.40	321.87	14.42

ecological environment, improve the landscape, cope with climate change, develop socio-economic factors, improve the quality of people's life and development, and contribute to the sustainable development of the country. Based on this policy guideline, Bac Lieu Province has concentrated on planting protective forests, which led to the rapidly increasing forest restoration rate from 2021 to 2022.

6.4 Deforestations from 2020 to 2022

The deforestation status shows the lowest loss in 2022 and similar high losses in both 2020 and 2021 (Table 7). The ratio of deforestation area in 2020, 6.66%, was more than double that of the forest restoration ratio in the same year. The situation became worse in 2021 with 14.96% of deforestation area comparing to 3.12% restoration rate, but in 2022, the situation improved slightly, with a deforestation rate of 14.42% versus a restoration rate of 6.47%.

Interestingly, developing intensive shrimp culture areas in Bac Lieu Province has led to the conversion of natural mangrove forests into shrimp ponds, especially in the 1990s [12]. Due to the high economic return, more farmers have started cultivating shrimp, and so they changed their land use from agricultural land and/or forest land into aquacultural land. So, despite multiple efforts from national, provincial, and local authorities, the protective forest area along the coastline is still in decline.

Figure 7 shows an important correlation between the restoration and destruction of protective forests, in which the most damage is seen from 2020 to 2021, and the years from 2021 to 2022 show a rapid increase in forest recovery (twice as much as in 2020–2021). At the same time, the area of deforestation also decreased. The difference ratio between the forest restoration area and deforestation area was the highest in 2021, but the decrease in 2022 shows the impact that Decision No. 524/QD-TTg has had. Despite all efforts by the government to restore and preserve the protective forests in the region, the total area of protective forest has decreased by about 20% in only 4 years.

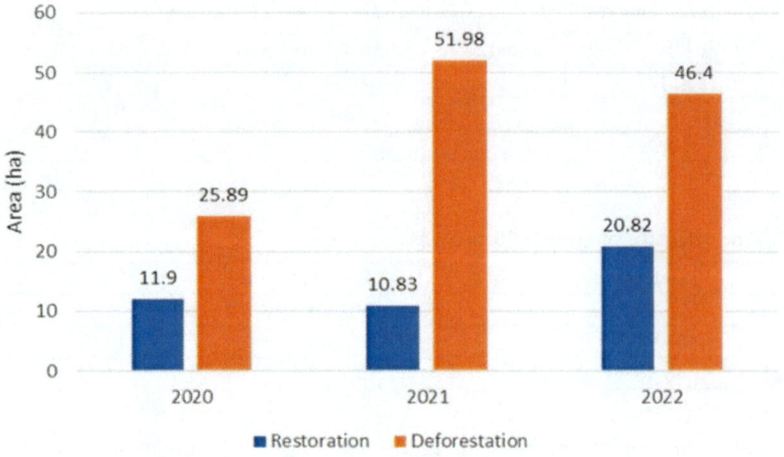

Fig. 7 A comparison between restoration area and deforestation area

7 Conclusion and Discussion

For this study, Sentinel-1 and Sentinel-2 images were integrated to detect changes in the protective coastal forest coverage in the Vietnamese Mekong Delta between 2019 and 2022. This is a new method used to try to overcome the shortcomings of optical data. A Random Forest classification approach was used to identify areas of forest loss and forest restoration. The Random Forest method has been shown to be effective in determining forested and non-forested areas, with accuracies of over 90%. The data for the years 2019–2022 show that during the three observed time periods (2019–2020, 2020–2021, and 2021–2022), the areas where new forest was detected ranged between 11.9 and 20.82 ha, while at the same time the areas of detected forest loss ranged between 25.89 and 51.98 ha. This means that there is a net loss of protective forest in this area, due to various factors. This result contributes to the assessment of the current status of the protective forest and the effectiveness of government projects on restoration and protection of protective forests in Bac Lieu Province.

References

1. Besset M, Anthony EJ, Brunier G, Dussouillez P (2016) Shore-line change of the Mekong River delta along the Southern part of the South China Sea coast using satellite image analysis (1973 – 2014). Géomorphologie Relief Proc Environ 22:137–146. https://doi.org/10.4000/geomorphologie.11336
2. Besset M, Gratiot N, Anthony EJ, Bouchette F, Goichot M, Marchesiello P (2019) Mangroves and shoreline erosion in the Mekong River delta, Viet Nam. Estuar Coast Shelf Sci 226:106263. https://doi.org/10.1016/j.ecss.2019.106263

3. Breiman L (2001) Random forests – machine learning, vol 45. Kluwer Academic Publishers, pp 5–32
4. Breiman L, Friedman J, Olshen R, Stone C (1984) Classification and regression trees. Wadsworth, Belmont
5. Criminisi A, Shotton J, Konukoglu E (2016) Decision forests for classification, regression, density estimation, manifold learning and semi-supervised learning Microsoft Research technical report TR-2011-114. https://www.microsoft.com/en-us/research/wp-content/uploads/2016/02/decisionForests_MSR_TR_2011_114.pdf
6. Groenewold S, Albers T, Sorgenfrei R (2019) Protect the coastal area of the Mekong Delta. Hanoi
7. Huynh HX, Loi TTT, Huynh TP, Tran VS, Nguyen TNT, Niculescu S (2019) Predicting of flooding in the Mekong Delta using satellite images. In: Vinh P, Rakib A (eds) Context-aware systems and applications, and nature of computation and communication. Lecture notes of Institute for Computer Sciences, Social Informatics and Telecommunications Engineering, vol 298. Springer, pp 143–156. https://doi.org/10.1007/978-3-030-34365-1_11
8. Huynh HX, Nguyen KM, Nguyen KD, Luong HH, Tran NC, Nguyen LTT, Tran TM, Pham PTT, Niculescu S (2020) Discovered changes in rice occupation with satellite images based on Random Forest approach. Proceedings of the 4th ACM International Conference on Machine Learning and Soft Computing (ICMLSC 2020). 86–97
9. Lappe R, Ullmann T, Bachofer F (2022) State of the Vietnamese coast – assessing three decades (1986 to 2021) of coastline dynamics using the Landsat archive. Remote Sens 14:2476. https://doi.org/10.3390/rs14102476
10. Lee J, Jurkevich L, Dewaele P, Wambacq P, Oosterlinck A (1994) Speckle filtering of synthetic aperture radar images: a review. Remote Sens Rev 8:313–340. https://doi.org/10.1080/02757259409532206
11. Marchesiello P, Nguyen NM, Gratiot N, Loisel H, Edward J. Anthony, Dinh CS, Nguyen T, Almar R, Kestenare E (2019) Erosion of the coastal Mekong delta: assessing natural against man induced processes. Cont Shelf Res 181: 72–89 https://doi.org/10.1016/j.csr.2019.05.004
12. Nam VN (2010) Inventory on the biodiversity of mangrove flora in order to find out which species thrive in particular environments and propose solutions for sustainable use and management of these coastal resources in Bac Lieu province. Project on Sustainable management of coastal forest ecosystems in Bac Lieu province. GTZ – Germany
13. Niculescu S, Lam NC (2019) Geographic object-based image analysis of changes in land cover in the coastal zones of the Red River Delta (Vietnam). J Environ Prot 10(3):413–430. https://doi.org/10.4236/jep.2019.103024
14. Pettorelli N (2013) The normalized difference vegetation index:1–208. https://doi.org/10.1093/acprof:osobl/9780199693160.001.0001
15. Ragia L, Pavlos K (2019) Monitoring the changes of the coastal areas using remote sensing data and geographic information systems. Proceeding of the Seventh International Conference on Remote Sensing & Geoinformation of the Environment (RSCy 2019). 111740X. https://doi.org/10.1117/12.2533659
16. Vietnamese Official Gazette (2021) Decision No. 524/QD-TTg approving the project title "Planting a billion trees from 2021 to 2025" dated on April 01st 2021 by Vietnamese Prime Minister. Vietnamese Official Gazette No. 543+544

Spatial Monitoring of Coastal Protection DikesCase Study of the Touristic Beach "Palm Beach, West Algiers, Algeria"

Walid Rabehi, Otmani Housseyn, Mohamed Amine Bouhlala, Sarah Kreri, Oussama Benabbou, Mohammed El Amin Larabi, and Hadjer Dellani

Abstract Seaside tourism is one of the most accessible summer activities for the population, especially on the coast of metropolitan cities. A typical example from the southern Mediterranean is Palm Beach, Algeria, one of the most populated beaches of the Algerian capital, where intense seasonal human traffic combined with the depletion of local sediments contributes to an intensification of coastal erosion. The entire beach was affected until the authorities started to build protective breakwaters. The objective of this work is, therefore, to establish a spatial monitoring of the dynamics of this beach, thanks to the Algerian Alsat2 satellite archives. The evaluation will be applied "before," "during," and "after" the construction using several spectral algorithms of detection of the shoreline. The monitoring will allow the quantitative observation of the behavior of the beach and thus to qualify the degrees of effectiveness of these breakwaters. It is also a question of determining the effectiveness of the high spatial resolution imagery (2.5 m) of this national optical sensor in the measurement of macro-erosion phenomena in order to generalize this approach to the entire Algerian coast in the future.

Keywords Erosion · Beach · Alsat2 · Breakwater · Remote sensing · GIS · Algeria

W. Rabehi (✉) · M. A. Bouhlala · S. Kreri · O. Benabbou · M. E. A. Larabi · H. Dellani
Centre des Techniques Spatiales, Algerian Space Agency, Arzew, Algeria
e-mail: wrabehi@cts.asal.dz; abouhlala@cts.asal.dz; skreri@cts.asal.dz; obenabbou@cts.asal.dz; mlarabi@cts.asal.dz; hdellani@cts.asal.dz

O. Housseyn
National School of Marine Sciences and Coastal Management – Campus Universitaire de Dely Ibrahim, Algiers, Algeria

S. Niculescu (ed.), *European Spatial Data for Coastal and Marine Remote Sensing*, https://doi.org/10.1007/978-3-031-16213-8_9

1 Introduction

The coastal environment is a complex ecosystem, composed of a multitude of natural factors that interact to maintain a stable balance [2]. This balance is permanently under stress and is often disturbed by several direct factors; anthropogenic, as urbanization, obstruction dams, sand theft or physical such hydrodynamic which is increasing by global climate change [13], but also indirect factors such as current modification caused frequently by inappropriate coastal defense facilities [30].

In Algeria, coastal erosion is a national issue and is more frequent on the sandy beach [4]. Moderate erosion to extreme erosion (causing complete extinction of beaches) [18] affects the whole 1600 km country coastlines, and as a reaction, several protection facilities have been established in touristic areas and large coastal cities [8, 28]. This coastal fringe is the most attractive area in Algeria and contains 60% of the Algerian population (in only 10% of the total Algerian area), demonstrating the socioeconomic pressures on this sensitive area [29].

Spatial monitoring of coastal areas by remote sensing techniques allows several possibilities of quantification of erosion/accretion balance [24, 33] and beach behavior especially before and after the establishment of coastal protection facilities (dikes, breakwaters, groynes); this allows to assess the quality of those protections and preconize alternative solution for decision-makers, for a better sustainability [10].

Coastline detection can be provided by several remote sensing techniques, meanly with optical images and aerial photography [34], starting with classical pixel classifier approaches (maximum likelihood, support vector machine, K-means) [35] showing some important radiometric confusion. Other approaches became more popular such as object-based approach [6] or more specifically filter-based methods applied in spectral break between land and sea [26]. Other experimental methods exist but are less popular such stereoscopy or using interferometric synthetic aperture radar data (InSAR) [32]. However, all those approaches even if they provide a time gain remain less precise than the Differential Global Positioning System (DGPS) coastline approach [24].

The more efficient optical data, widely used for coastline monitoring, are high resolution data such QuickBird/Ikonos data [11], aerial photography [34], or more recently Pleiades images, which has a spatial resolution lower than 1 m allowing to assess the micro-erosion (such as seasonal shoreline dynamics), but remain economically expensive especially for developing countries such Algeria where research funding is limited.

The Algerian Space Agency through the National Spatial Program (2006–2020) detains two optical satellites in orbit "Alsat 2A and Alsat 2B" [15], which provide scenes with 2.5 m resolution, lower than QuickBird/Pleiades commercial satellites (which can assess micro-erosion), but allow to detect macro-erosion case of large beaches, which is a frequent phenomenon in Algerian coasts (large erosion case loss, storm impacts, beach extinction).

The aim of this work is to perform a spatial assessment of beach dynamics in the west of the Algerian capital (the touristic beach "Palm Beach/west Algiers"), using specific spectral filtering approaches (basing on the near-infrared insensitivity) never experimented in the south Mediterranean shore. This beach is under extreme human pressure and has been protected by fives dikes in 2013, as it undergoes high erosion episodes [3] in the last year, and faced in the last decade a risk of complete extinction.

This work will provide a spatial coastline monitoring (before and after protection) using the Algerian satellite Alsat 2, in order to assess this "macro-erosion," and will contribute to evaluate the efficiency of this spatial sensor (2.5 m) compared to classical in situ tools such as DGPS campaign, which is more precise but with a higher cost (time/budget) [6].

2 Study Area

The area of interest is located in the west of Algiers in the bay of Bousmail, between the Sidi Fredj touristic port and the Sable d'Or beach in Zéralda [5]. By the term "Palm Beach," we also include Azur Beach in the west (due to its proximity and its natural continuity with our interest area); the area is also limited by two rivers that are both arms of the Mazafran River (Fig. 1).

After the high erosion episodes (2006–2011), local authorities started building six marine dikes (Fig. 2) with the following characteristics:

- Width of the central concrete structure: 6 m
- Length of the central concrete structure: 95 m
- Width of riprap surface: 3–5 m
- Size of composite rocks: 1–2 m
- Height of the structure: emergent, 1–1.5; submerged, 1 m

Later, those dikes start creating a new sediment accumulation area (called Tombolo, or sediment barrier), which is assessed in this study (Fig. 3).

3 Methodology

The pre-processing of the Alsat 2 optical images included pansharpening and radiometric correction to produce an optimal reflectance pixels of 2.5 m resolution. Image analysis included shoreline extraction using several spectral approaches (supervised spectral angle matching index and morphological mean filter) and unsupervised using principal component analysis (PCA)/independent component analysis (ICA) and normalized difference water index (NDWI). Once the water–land classes are identified, the wet pixel is analyzed using an unmixing algorithm to

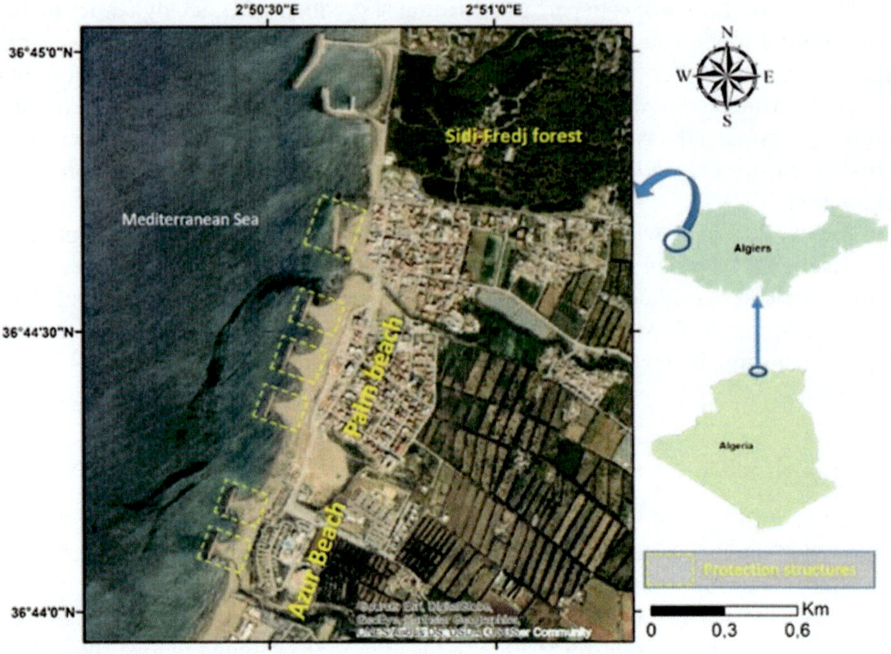

Fig. 1 Study area (West Algiers, Algeria)

Fig. 2 Dikes' structure

Fig. 3 New extended beach (sand Tombolo)

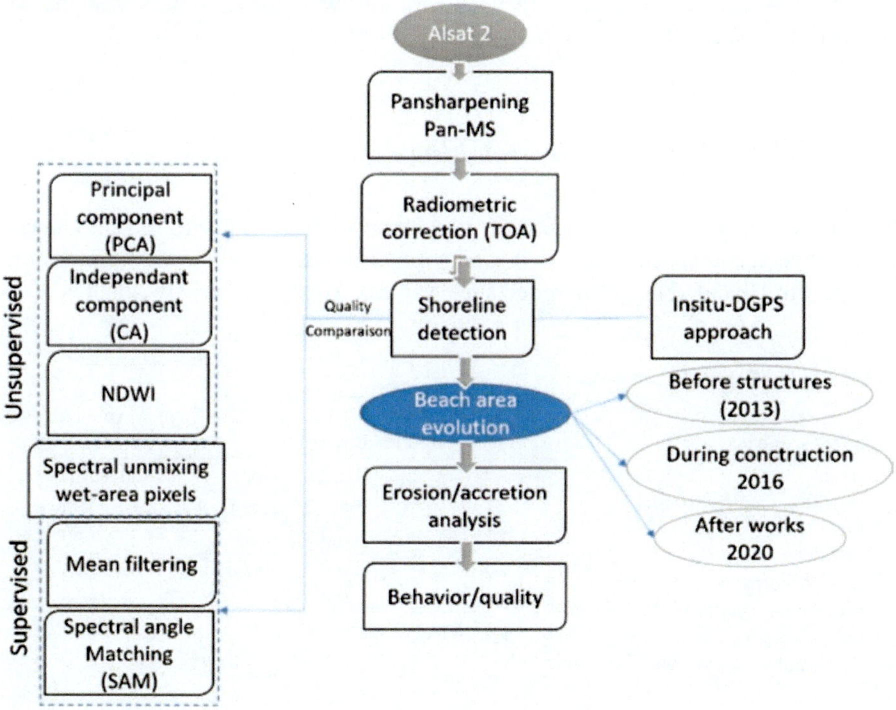

Fig. 4 Overall methodology

identify clearly the shoreline and delineate the beach area. The quality of this shoreline product is compared to a DGPS reference shoreline in order to obtain the overall accuracy (Fig. 4).

Once the best algorithm is identified (after assessing all the approaches individually), we produced multitemporal evolution of the beach area between 2013 (before protection) and 2014 in the middle of work protection and in the end 2021 (after the work's end "state of stability").

3.1 Data Set and Pre-processing

In this study, Alsat 2 images were used with multispectral (10 m) and panchromatic (2.5 m) bands. The characteristics are presented in Table 1.

Before being corrected, the image of Alsat 2A encoded in 10 bits is converted into luminance (radiance) values [9] expressed in W/(m^2 × sr × μm), by applying the following calibration equation:

$$L = DN.g^{(-1)} + L\_0 \tag{1}$$

with *L* as the radiance, *DN* as the digital numbering, and *g* and *L_0* as calibration coefficient (gain and bias).

Then, in the second step, the scene is converted to reflectance value using the FLAASH module [19] on the ENVI 5.6 © software using a filter specific to NAOMI-1 sensor.

3.2 SAM: Spectral Angle Matching

The spectral angle is used because of its spectral rupture ability (best radiometric ratio between two spectral bands).It is an n dimension angle where each dimension is a spectral band (based on reflectance values). If the angles are smaller, this

Table 1 Alsat 2 data characteristics

Satellite scene	Resolution	Bands	Wavelength (nm)
Alsat 2A Acquisition date: 11/10/2013 27/09/2014 29/01/2021 Source: Algerian Space Agency (ASAL) Elementary scene: 17,5 × 17,5 Km (306,25 km^2) Revisit time: 3 days with 30° angle	10 m	Band 1 – blue	450 ± 520
		Band 2 – green	530 ± 590
		Band 3 – Rred	620 ± 690
		Band 4 – Nnear infrared	760 ± 890
	2.5 m	Panchromatic	450 ± 745

indicates a closer match to the reference band [20]; the filter produced is mainly based on the near-infrared band and its combination with other bands.

3.3 Median Filter

The median filter is a non-linear morphological filter, which allows a better preservation of the edges of kernel filters. The filter acts by ranking all pixel reflectance values in the window and then takes the median values (instead of the mean) [12] for a better identification of the spectral breaks and class separability.

3.4 Spectral Unmixing

Typical spectral unmixing techniques aim at unsupervised extracting, from pixels considered optical remote sensing data, pure material (also called endmember) spectra, and inferring the proportion (also called abundance fraction) of each of these spectra in each pixel of that data [7].

In this contribution, an unsupervised linear spectral unmixing technique, by means of the standard multiplicative nonnegative matrix factorization [22] technique initialized by the vertex component analysis [7] approach, is used to unmix the considered multispectral data only on the wet area that corresponds to the contact "water–land" zone. This unsupervised unmixing process is feasible and relevant [16] since the objective here is to unsupervisedly extract only two endmember (water and land) spectra and their abundance fractions from the used four spectral bands of the considered multispectral data. Then, the shoreline is estimated by analyzing obtained abundance fractions of the two considered endmembers.

3.5 Principal Component Analysis (PCA)

Principal component analysis (PCA) is a classical approach of dimensionality reduction [27]. Its efficiency is widely proved for noise elimination in remote sensing [31]. It provides for remote sensing classification and orthogonal axes who demonstrate the uncorrelated bands and where the spectral variance is the key element, to provide the best radiometric separability.

3.6 Independent Component Analysis (ICA)

Independent component analysis (ICA) is an efficient approach of empirical data separation, based on the non-Gaussian assumption of the independent data sources. No prior information is needed. IC transformation can distinguish features of interest even when they occupy only a small portion of the pixels in the image [17].

In contrast to correlation-based transformations such as principal component analysis (PCA), ICA not only decorrelates the signals (second-order statistics) but also reduces higher-order statistical dependencies [21].

3.7 Normalized Difference Water Index (NDWI)

The normalized difference water index (NDWI) was developed for the extraction of water features from Landsat imagery [23] using near infrared and blue and is very efficient in natural landscapes (forests, natural coasts) but is judged less efficiently in the artificial coastal landscape (radiometric confusion between water and artificial classes such as bitumen) [1, 14].

4 Results

The land–sea transect analysis in the derived image (Fig. 5a) produced by the mean filter algorithm shows significant variability (increasing from land to sea). The anchorage area is clearly marked by a clear radiometric rupture (Fig. 5b).

The other indicator generated from the supervised approach is the spectral angle, which demonstrates tendencies opposite to the previous index (MF), with more positive on land and decreases going offshore, and the break is present but less marked than the previous index (Fig. 6).

For the ratio between the two previous indicators, the general overview displays a lot of spectral noise "heterogeneity" (Fig. 7a). However, the morphological index MF provides more influence to the result (slight growth while going out to the sea), but the break "land–sea" is, however, less marked than the two independent spectral indexes (Fig. 7b).

Spectral unmixing provides acceptable quality for pixels where abundance is up to 70% but still has some difficulties with the turbidity area, which, however, does not obstruct the extraction of a relatively correct shoreline (Fig. 8).

Regarding the unsupervised approaches and without any prior user learning, the NDWI spectral index displays a clear discrimination between beach and sea (Fig. 9a), and even if this image contains confusion with urban objects, the index provides one of the best results for the coastline, reinforced by the spectral break (see the diagram of spectral profile, Fig. 9b).

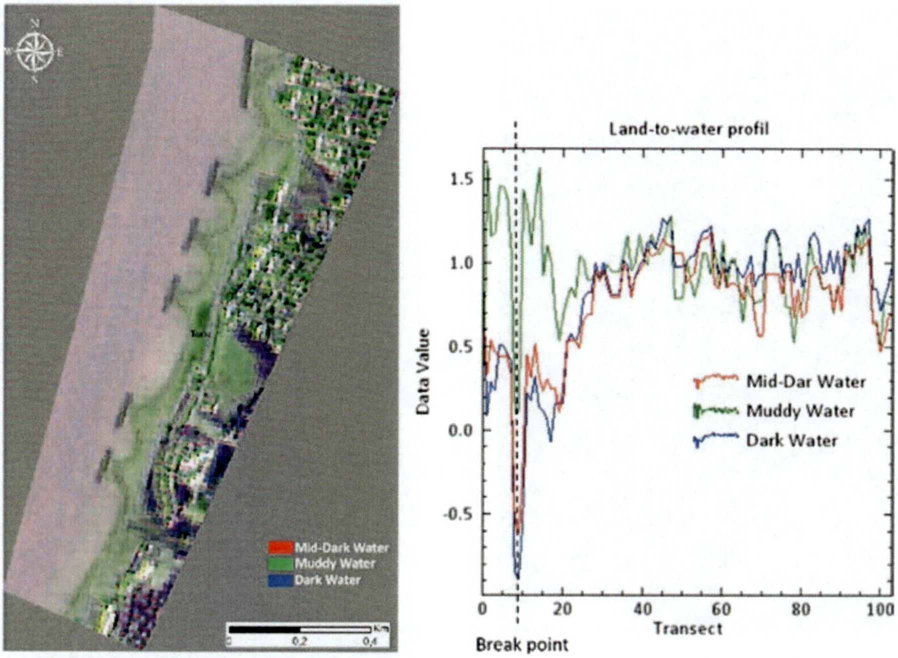

Fig. 5 Mean filter map (**a**) and spectral transect (**b**)

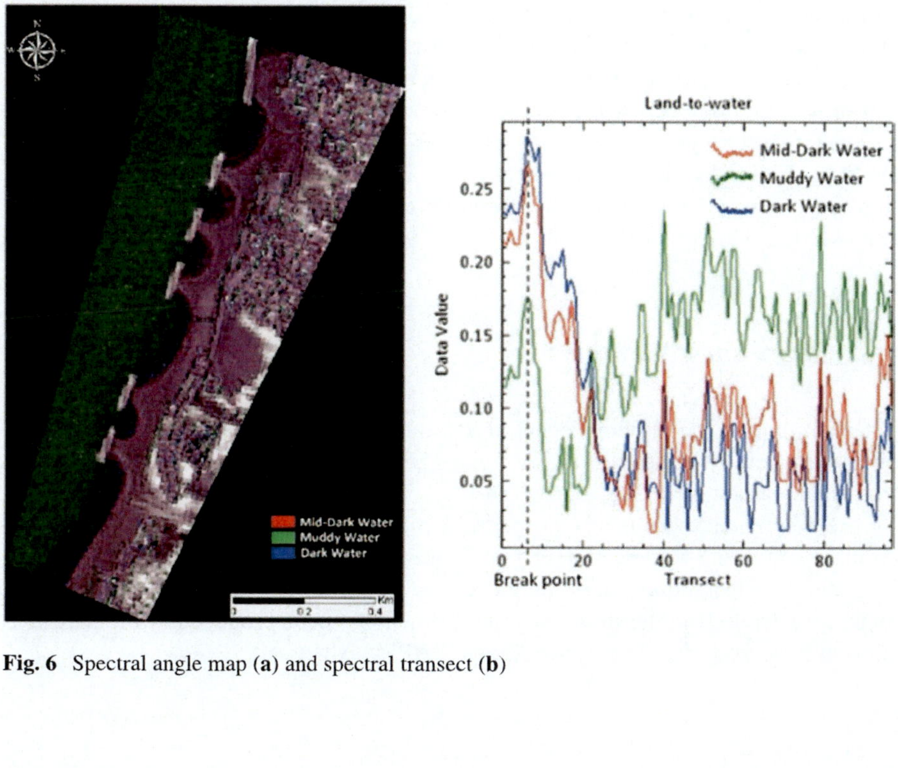

Fig. 6 Spectral angle map (**a**) and spectral transect (**b**)

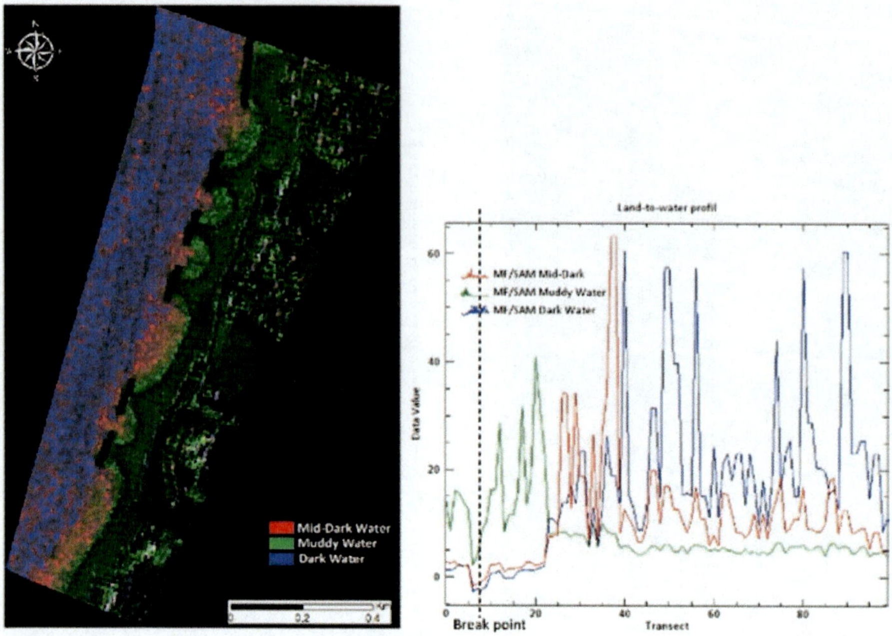

Fig. 7 Ratio "SAM/MF" map (**a**) and spectral transect (**b**)

For the principal component analysis (Fig. 10a), it is the first correlation axis that allows the best land–sea difference (Fig. 10b); it thus groups together the majority of statistical variances, and there is, however, a clear visual absence of sensitivity toward water turbidity (muddy water, clear water, etc.).

For the last approach, which is, like the previous one, an effective approach of dimensionality reduction, the land–sea break is marked with a lower quality than the PCA (Fig. 11a) with also a kind of confusion between the sea and the bare interior soils. The second and third correlation axis of the ICA displays the clearest land–sea breaks (Fig. 12b).

4.1 *Shorelines' Variants*

After computing the previous spectral variants in order to produce the best visual distinction between land and sea, shorelines are created manually as a GIS Shapefile from each variant map (Fig. 12).

The reference shoreline is produced by ground surveys using a Differential Global Positioning System (DGPS) using the real-time kinematic (RTK) approach [24], and this technique offers the most precise tool of shoreline detection (centimetric precision) [6]. The shoreline measurement has been produced by the Algerian Littoral Promotion and Protection Agency (APPL-Algeria).

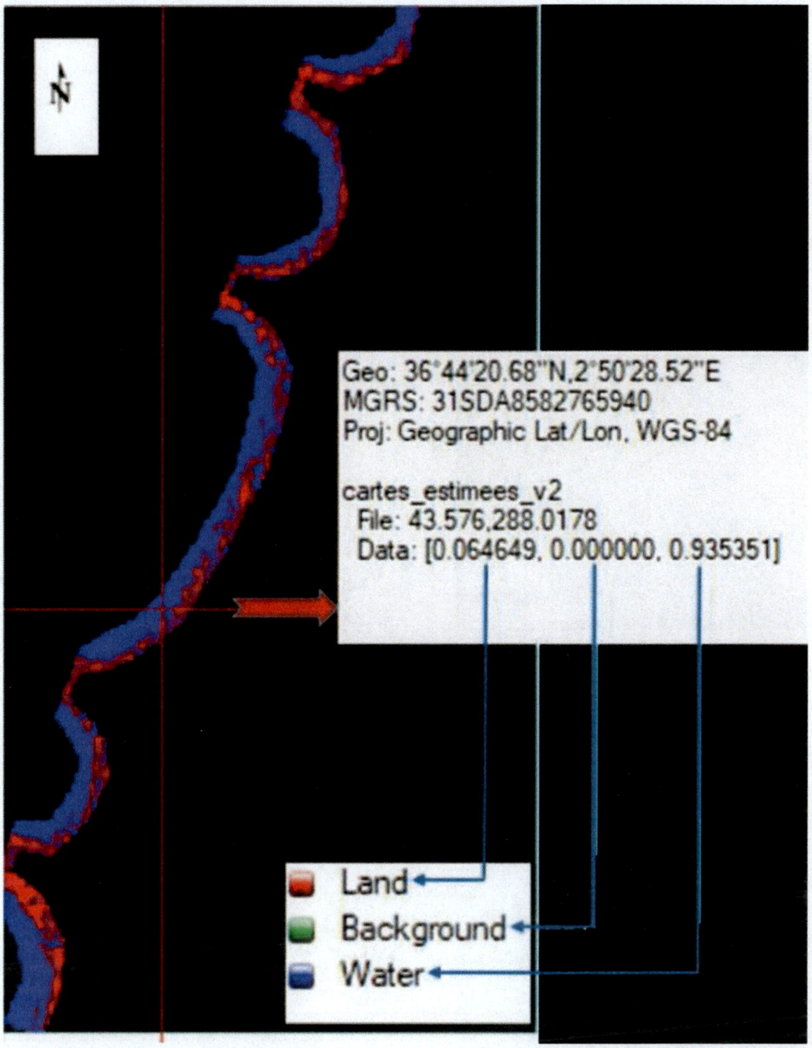

Fig. 8 Spectral unmixing abundance/endmember map

Considering the spatial sensitivity of the Alsat 2 scene (2.5 m × 2.5 m) and for the accuracy assessment purposes, a buffer area is produced from each shoreline vector file and clipped with the reference shoreline. The overall accuracy statistics are synthesized in Fig. 13 below.

The principal component analysis is the classifier that provides the best accuracy followed by the NDWI spectral index. Furthermore, the spectral angle mapper provides the lowest accuracy values, and this is due to their excessive sensitivity of muddy water in the "surf zone."

Most of the confusion over the exact location of the surf zone is due to the turbidity of this contact zone but also the coarseness of the pixel (2.5 m × 2.5 m); however,

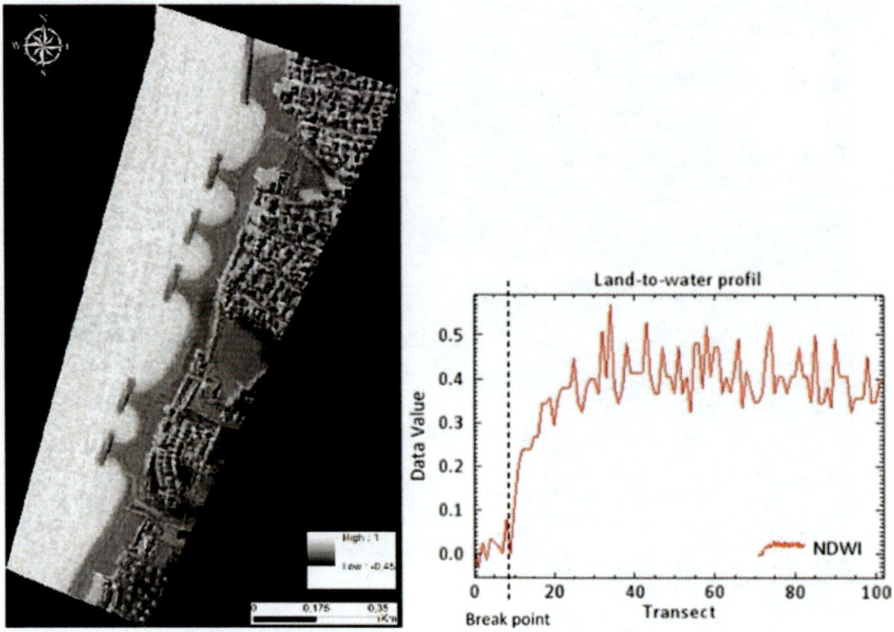

Fig. 9 NDWI map (**a**) and spectral transect (**b**)

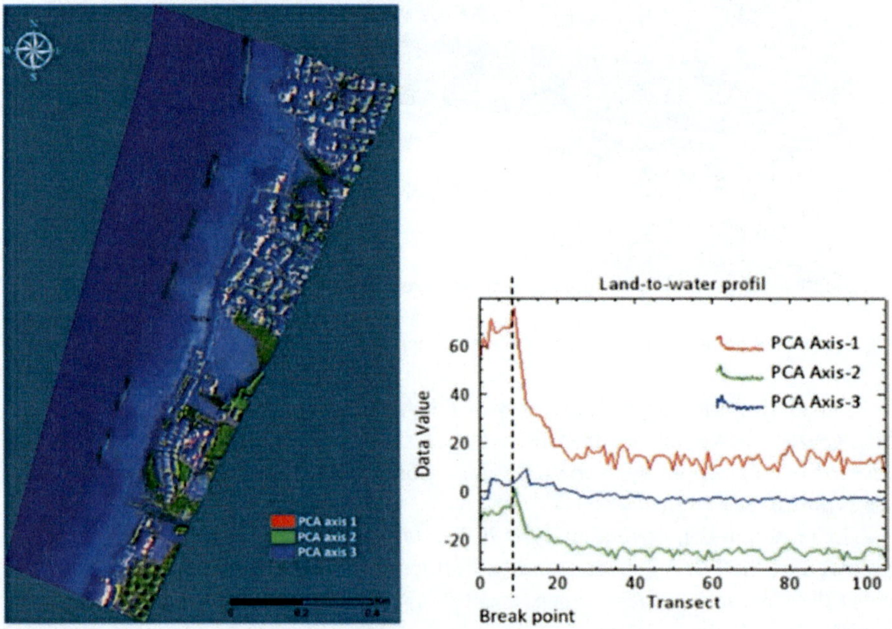

Fig. 10 PCA map (**a**) and spectral transect (**b**)

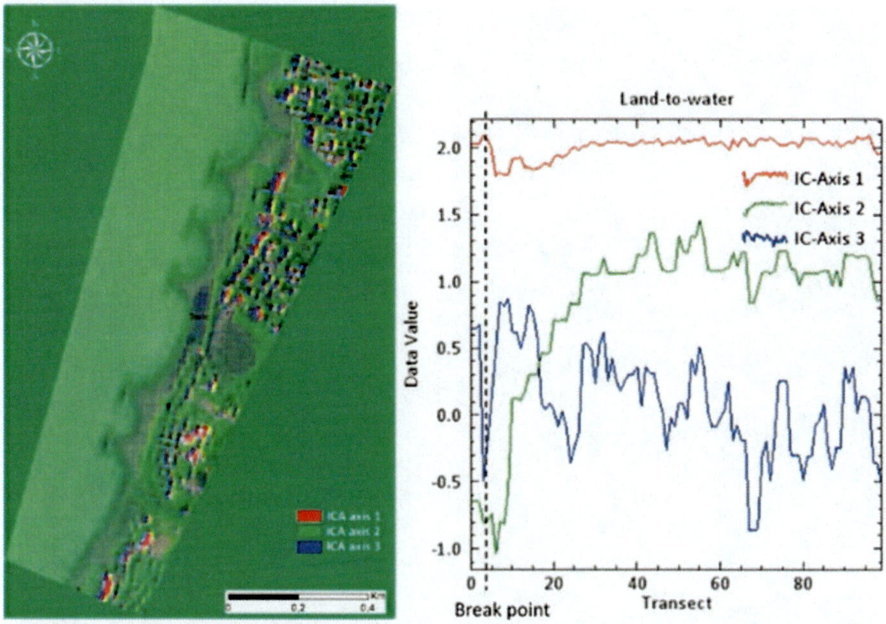

Fig. 11 ICA map (**a**) and spectral transect (**b**)

even if it is identical, the different shorelines are similar in the path shape of the line on the majority of the beach (Fig. 14).

In a specific location of the beach, the ICA particularly confused the turbid water (Fig. 15a), and this location (proximity of the last protection structure) is a false land positive by ICA approach (Fig. 15b).

The statistical analysis of spectral break variability was computed from the three proximity pixels located in the break area (on either side of water and land), using a mathematical subtraction of the reflectance values (Fig. 16).

The result matches partially with the accuracy results where PCA and NDWI approaches are the ones that provide the best spectral break "land–water."

4.2 Beach Dynamic Analysis

The PCA provides the best shoreline extraction from the optical Alsat 2 images; thus, the surface of the beaches was digitized (Fig. 17). Table 5 below shows also the area statistics for the beach evolution (Table 2).

The before-construction period (2013) was characterized by a relatively small area (4.38 hectares), reflecting the very vulnerable situation of the beach (in constant erosion for over 10 years).

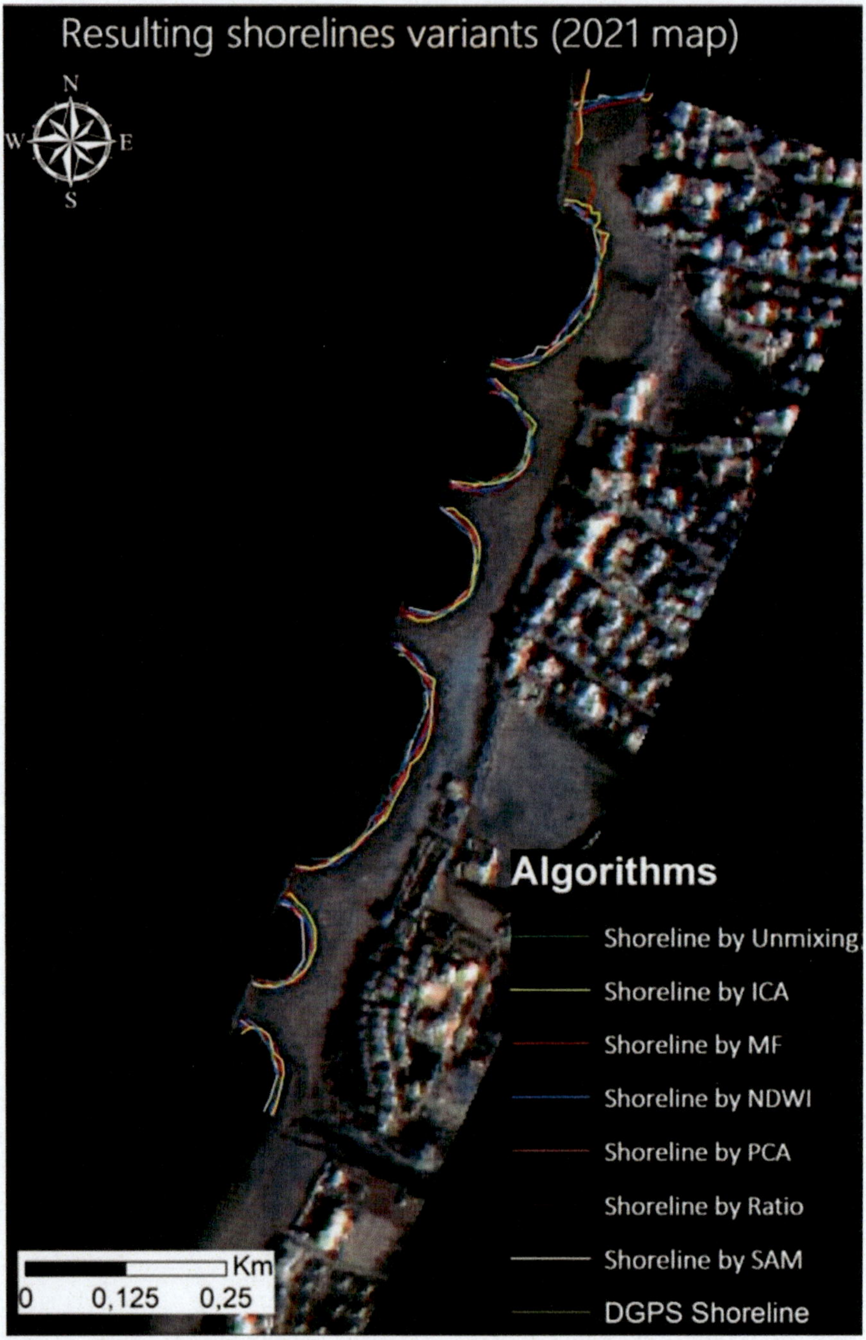

Fig. 12 Shorelines' comparison

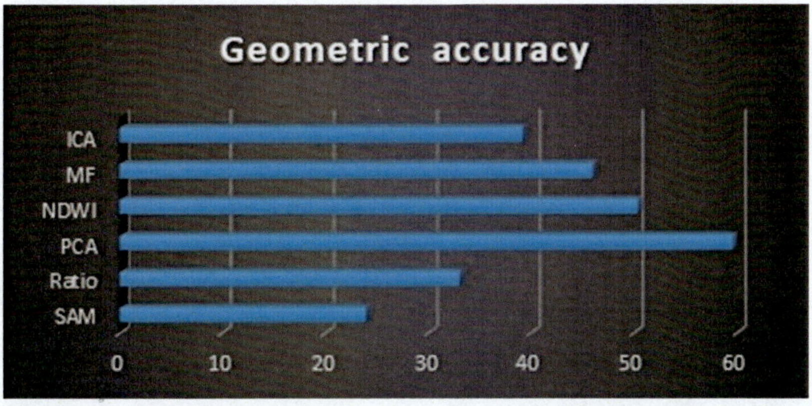

Fig. 13 Shorelines' accuracy

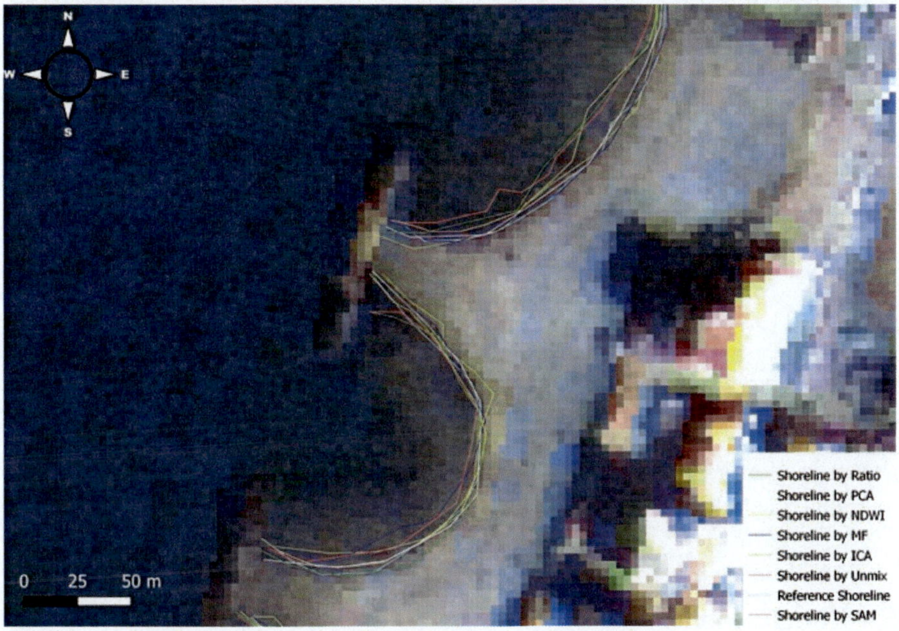

Fig. 14 Global path similarity

The year 2014 (start of works) demonstrates a slight increase, but which is not necessarily linked to the start of the beach stabilization but rather to the contribution of artificial sediment and solid deposits of materials (sand, gravel, rocks) and also to the decrease in touristic attendance of the beach due to the construction works.

The year 2021 (5 years after the end of the works) displays a very strong increase in the surface of the beach (+ 400%) where the effects of current attenuation by the

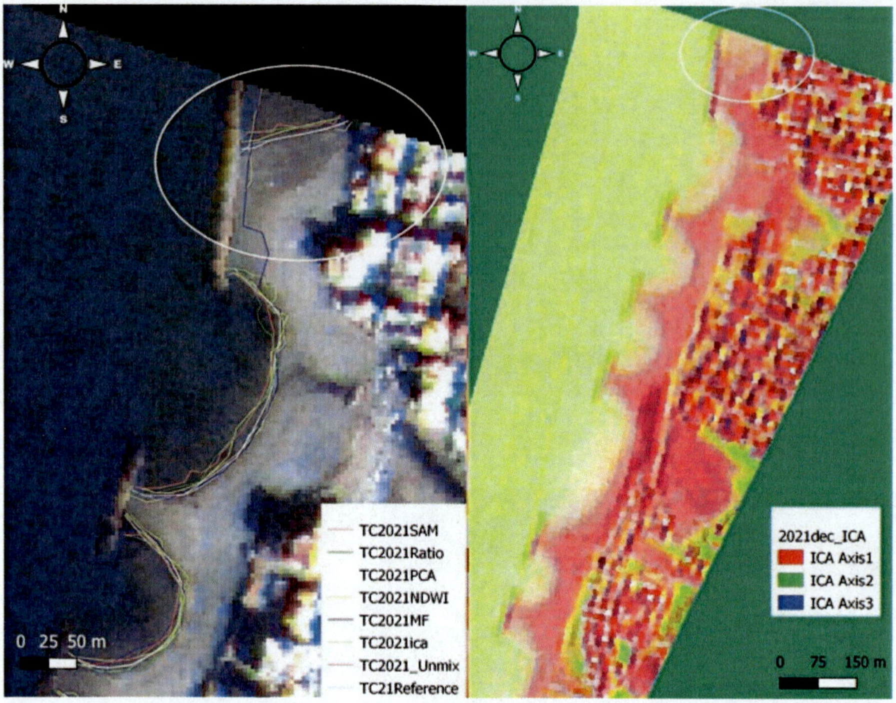

Fig. 15 Area of total discord: (**a**) Alsat2 image and (**b**) ICA index

Table 2 Area evolution of the beach

	2013	2014	2021
Beach area (hectares)	43,805	48,205	22,9606

new protection structures (as well as the artificial sediment supply) have contributed to a significant gain in beach area and high relative stability.

5 Conclusion

The present analysis demonstrates a remarkable recovery of the beach after the installation of the protection structures confirming the correct analysis of the hydrodynamic context; however, the performed monitoring does not take into account the exceptional erosion episodes (storm, seasonal erosion).

Supervised approaches show relatively poor results (SAM, MF, ratio), which are probably due to the sampling process that tends to generalize the significant spectral variability; we can also point to the coarseness of the spatial resolution of the pixel

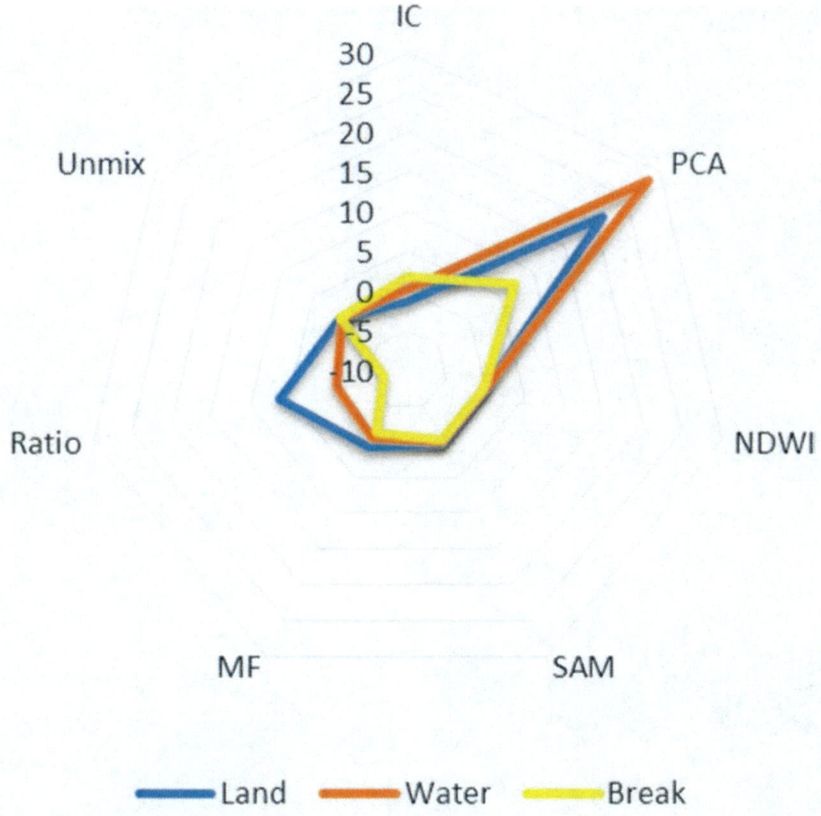

Fig. 16 Spectral break diagram (land to water)

(2.5 m), which has quite brutal spectral breaks; we also note that the turbidity of the breaking wave zone at the shore also produces a considerable spectral noise.

Spectral unmixing even not widely used for multispectral data had provided in this study through the pixel endmembering and abundance acceptable results for the radiometric separation between land and water in the wet area closely compared to the two most efficient approaches (PCA and NDWI).

The non-exhaustiveness of the tested algorithms in this contribution is also linked to the choice of a simplistic and reproducible indicator on a large scale (quest for future generalization) but also to the local state of the art where certain methods have already been applied (object-based segmentation, SVM, Random Forest classifiers) but on a different spatial resolution.

The methodology can also be improved in a next phase in combination with multisource data, in particular, active sensor radar and the technical possibility of producing Alsat 2 stereoscopic scenes by producing a digital elevation model to provide a new discriminating information in relation with the beach-end bank.

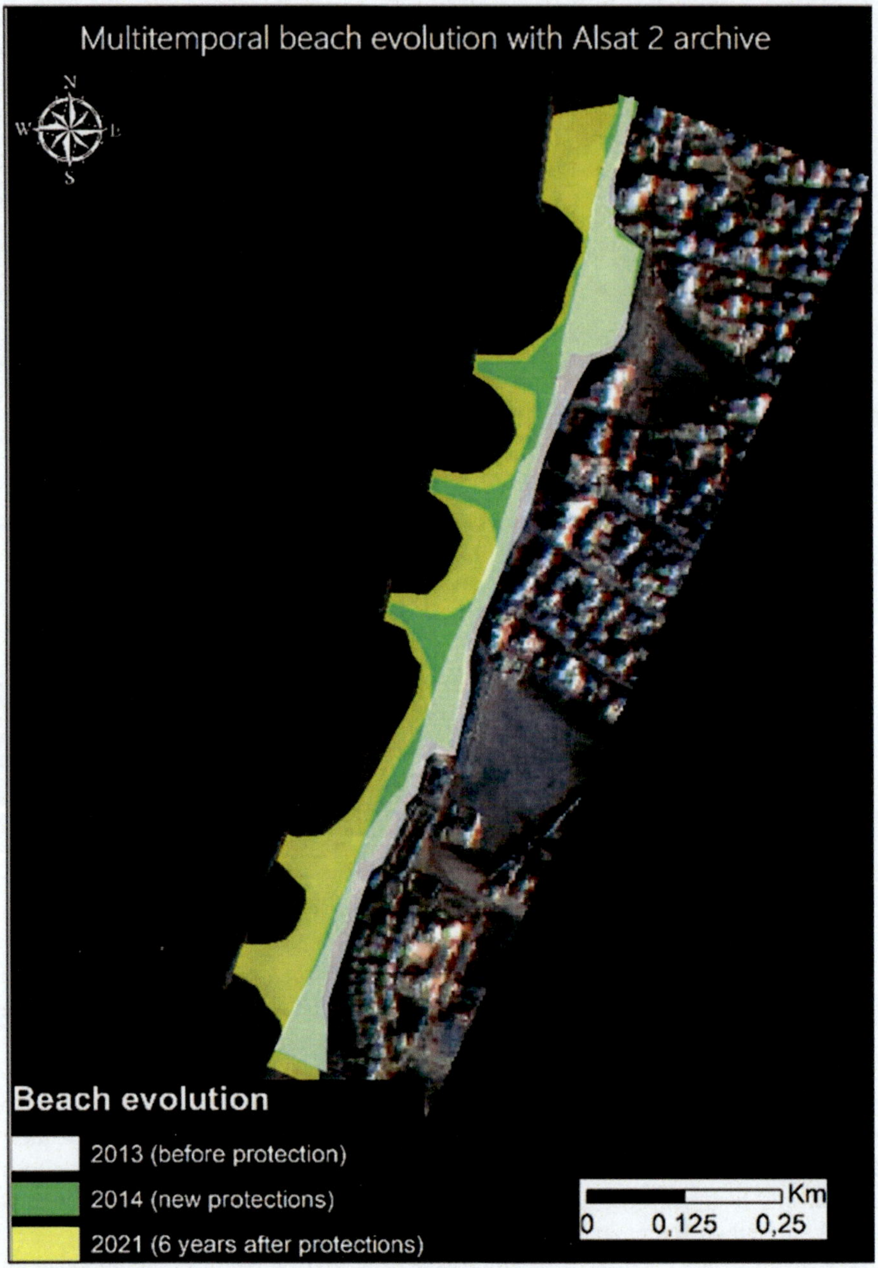

Fig. 17 Multitemporal evolution of the beach

Overall, the adopted approach can be the subject of an effective decision-support tool for the integrated management of the coastline and the local actors, in particular, through the deployment of Alsat imagery for the entire Algerian coastline.

Acknowledgments This work is part of a research project titled "Spatial land use indicators of the Algerian coastal zone" operated by the Centre des Techniques Spatiales/Algerian Space Agency (Arzew, Algeria), with the contribution of the National High School of Marine Science and Coastal Management (ENSSMAL Algiers, Algeria).

References

1. Adam E, Mutanga O, Rugege D (2010) Multispectral and hyperspectral remote sensing for identification and mapping of wetland vegetation: a review. Wetl Ecol Manag 18(3):281–296. https://doi.org/10.1007/s11273-009-9169-z
2. Al-Awadhi T, Mansour S, Hereher M (2020) Assessment of coastal sensitivity to non-eustatic sea level rise: a case study on Muscat coast—Sultanate of Oman. Arab J Geosci 13(10):371. https://doi.org/10.1007/s12517-020-05321-x
3. Al Sid Chikh S, Belkessa R (2011) Etude de protection contre l'érosion de la plage de Palm Beach-Azur, commune de Zéralda. Alger Recours aux modèles réduits physiques à fonds fixe et mobile:1–4. https://doi.org/10.5150/cmcm.2011.001
4. Amarouche K, Akpınar A, Çakmak RE, Houma F, Bachari NEI (2020) Assessment of storm events along the Algiers coast and their potential impacts. Ocean Eng 210(June):107432. https://doi.org/10.1016/j.oceaneng.2020.107432
5. Balaska A, & Benhaik K (2018) Vulnerability assessment of Palm beach coastal area Analyse de la vulnerabilité cotiére de plage de palm beach [ENSSMAL-Algiers]. http://library1.nida.ac.th/termpaper6/sd/2554/19755.pdf
6. Bengoufa S, Niculescu S, Mihoubi M. K, Belkessa R, Rami A, Rabehi W, & Abbad K (2021) Machine learning and shoreline monitoring using optical satellite images: case study of the Mostaganem shoreline, Algeria. J Appl Remote Sens, 15(2), 1–25. https://doi.org/10.1117/1.JRS.15.026509
7. Bioucas-Dias J.M, Plaza A, Dobigeon N, Parente M, Du Q, Gader P, Chanusso J (2012) Hyperspectral unmixing overview: geometrical, statistical, and sparse regression-based approaches. IEEE J Sel Top Appl Earth Obs Remote Sens 5, 354–379, doi: https://doi.org/10.1109/JSTARS.2012.2194696
8. Chaib W, Guerfi M, Hemdane Y (2020) Evaluation of coastal vulnerability and exposure to erosion and submersion risks in Bou Ismail Bay (Algeria) using the coastal risk index (CRI). Arab J Geosci 13(11):420. https://doi.org/10.1007/s12517-020-05407-6
9. Chander G, Markham BL, Helder DL (2009) Summary of current radiometric calibration coefficients for Landsat MSS, TM, ETM+, and EO-1 ALI sensors. Remote Sens Environ 113(5):893–903
10. Djouder F, Boutiba M (2017) Vulnerability assessment of coastal areas to sea level rise from the physical and socioeconomic parameters: case of the Gulf Coast of Bejaia, Algeria. Arab J Geosci 10(14). https://doi.org/10.1007/s12517-017-3062-5
11. Giannini M. B, & Parente C (2015) An object based approach for coastline extraction from Quickbird multispectral images. Int J Eng Technol, 6(6), 2698–2704
12. Haralick RM, Sternberg SR, Zhuang X (1987) Image analysis using mathematical morphology. IEEE Trans Pattern Anal Mach Intell 4:532–550
13. IPCC (2019) The ocean and cryosphere in a changing climate (Issue September). https://www.ipcc.ch/srocc/download/

14. Kafrawy S, Basiouny M, Ghanem E, Taha A (2017) Performance evaluation of shoreline extraction methods based on remote sensing data. J Geography Environ Earth Sci Int 11(4):1–18. https://doi.org/10.9734/jgeesi/2017/36233
15. Kameche M, Gicquel AH, Joalland D (2011) ALSAT-2A transfer and first year operations. J Aerosp Eng 3(2):67
16. Karoui MS, Deville Y, Hosseini S, Ouamri A (2012) Blind spatial unmixing of multispectral images: new methods combining sparse component analysis, clustering and non-negativity constraints. Pattern Recogn 45:4263–4278. https://doi.org/10.1016/j.patcog.2012.05.008
17. Kemker R, Salvaggio C, Kanan C (2018) Algorithms for semantic segmentation of multispectral remote sensing imagery using deep learning. ISPRS J Photogramm Remote Sens 145:60–77. https://doi.org/10.1016/j.isprsjprs.2018.04.014
18. Kermani S, Boutiba M, Guendouz M, Guettouche MS, Khelfani D (2016) Detection and analysis of shoreline changes using geospatial tools and automatic computation: case of jijelian sandy coast (East Algeria). Ocean Coast Manag 132:46–58. https://doi.org/10.1016/j.ocecoaman.2016.08.010
19. Khanh Ni TN, Tin HC, Thach VT, Jamet C, Saizen I (2020) Mapping submerged aquatic vegetation along the central Vietnamese coast using multi-source remote sensing. ISPRS Int J Geo Inf 9(6). https://doi.org/10.3390/ijgi9060395
20. Kruse FA, Lefkoff AB, Boardman JW, Heidebrecht KB, Shapiro AT, Barloon PJ, Goetz AFH (1993) The spectral image processing system (SIPS)—interactive visualization and analysis of imaging spectrometer data. Remote Sens Environ 44(2):145–163. https://doi.org/10.1016/0034-4257(93)90013-N
21. Lee TW (1998) Independent component analysis. In: Independent component analysis. pp 27–66. Springer. https://link.springer.com/chapter/10.1007/978-1-4757-2851-4_2
22. Lee DD, Seung HS (1999) Learning the parts of objects by non-negative matrix factorization. Nature 401:788–791. https://doi.org/10.1038/44565
23. McFeeters SK (1996) The use of the Normalised Difference Water Index (NDWI) in the delineation of open water features. Int J Remote Sens 17(7):1425–1432. https://doi.org/10.1080/01431169608948714
24. Otmani H, Belkessa R, Bengoufa S, Boukhediche W, Djerrai N, Abbad K (2020) Assessment of shoreline dynamics on the Eastern Coast of Algiers (Algeria): a spatiotemporal analysis using in situ measurements and geospatial tools. Arab J Geosci 13(3). https://doi.org/10.1007/s12517-020-5069-6
25. Otmani H, Belkessa R, Rabehi W, Guerfi M, Boukhdiche W (2019) Dégradation des dunes côtières algéroises entre pression de l'urbanisation et conséquences sur l'évolution de la ligne de rivage. Geo-Eco-Marina 2019(25):131–145. http://journal.geoecomar.ro/geo-eco-marina/article/view/10_2019
26. Parker B (2003) The difficulties in measuring a consistently defined shoreline – the problem of vertical referencing. J Coast Res 38:44–56
27. Pearson K (1901) Principal components analysis. The London, Edinburgh, and Dublin Philosophical Magazine and Journal of Science 6(2):559
28. Rabehi W, Guerfi M, Mahi H (2018) Cartographie de la vulnérabilité des communes de la baie d'Alger. Approche socio-économique et physique de la côte. Méditerranée. Revue Géographique Des Pays Méditerranéens/J Mediterranean Geography, May
29. Rabehi W, Guerfi M, Mahi H, Rojas-Garcia E (2019) Spatiotemporal monitoring of coastal urbanization dynamics: case study of Algiers' bay, Algeria. J Indian Soc Remote Sensing 47(11):1917–1936. https://doi.org/10.1007/s12524-019-01037-y
30. Rangel-Buitrago NG, Anfuso G, Williams AT (2015) Coastal erosion along the Caribbean coast of Colombia: magnitudes, causes and management. Ocean & Coastal Manag 114:129–144
31. Richards JA (1999) Remote sensing digital image analysis, vol 3. Springer
32. Salameh E, Almar R, Baptista P, Heygster G, Lubac B, Raucoules D, Almeida LP, Bergsma EWJ, Capo S, De Michele M, Idier D, Li Z, Marieu V, Poupardin A, Silva PA, Turki I, Laignel

B (2019) Monitoring beach topography and nearshore bathymetry using Spaceborne remote sensing, vol 11

33. Salem Cherif Y, Mezouar K, Guerfi M, Sallaye M, Dahmani AEA (2019) Nearshore hydrodynamics and sediment transport processes along the sandy coast of Boumerdes, Algeria. Arab J Geosci 12(24). https://doi.org/10.1007/s12517-019-4981-0
34. Sallaye M, Mezouar K, Salem Cherif Y, Dahmani AEA (2018) Morphological evolution of center Boumerdes in Zemmouri Bay (Algeria) from 1922 to 2017. Arab J Geosci 11(19):602. https://doi.org/10.1007/s12517-018-3950-3
35. Toure S, Diop O, Kpalma K, Maiga AS (2019) Shoreline detection using optical remote sensing. Int J Geo-Inform Rev 8(75):2–21. https://doi.org/10.3390/ijgi8020075

Monitoring Shoreline Changes in the Vietnamese Mekong Delta Coastal Zone Using Satellite Images and Wave Reduction Structures

Tran Van Ty, Dinh Van Duy, Huynh Thi Cam Hong, Nguyen Dinh Giang Nam, Huynh Vuong Thu Minh, Lam Van Thinh, Nguyen Vo Chau Ngan, and Nguyen Hieu Trung

Abstract This study aimed to assess the current state of the shoreline and the effects of erosion on the shoreline in Vinh Chau Town, Soc Trang Province, Vietnam. Satellite image overlays were used to quantify the variation in the shoreline as a result of erosion caused by changes in wave action. In addition, the wave measurements were implemented at three representative shoreline protection sections (sea dike, mangrove forest belt, and the breakwater) to evaluate the wave height reduction at the shoreline. The results showed that erosion affected approximately 23 km (32%) of 72 km coastline of the study area. The erosion penetrates the land area from −16.9 to −3.0 m/year (Landsat images) and −11.68 to −7.95 m/year (Google Earth images); the coastline recession increases every year, leading to the gradual loss of mangrove forests and also farmland. The wave measurement shows the effectiveness in wave height reduction of the mangrove forest and the constructed breakwater to protect the sea dike. Wave height reduces more than 50% when passing through the mangrove forest belt, corresponding to a maximum height (H_{max}) of 62.3%, 1/10 H_{max} at 55.3%, and 1/3 H_{max} at 54%. For the constructed breakwater, the wave reduction efficiencies recorded due to Hmean are 72.18% and 1/10 H_{max} are 73.16% and reach 72.47% with 1/3 H_{max}. The results are based on wave measurement over a short time period; thus, it is not possible to conclude about the wave

T. Van Ty · D. Van Duy · H. T. C. Hong
Department of Hydraulic Engineering, Can Tho University, Can Tho City, Vietnam
e-mail: tvty@ctu.edu.vn; dvduy@ctu.edu.vn; htchong@ctu.edu.vn

N. D. G. Nam · H. V. T. Minh · L. Van Thinh · N. V. C. Ngan (✉)
Department of Water Resources, Can Tho University, Can Tho City, Vietnam
e-mail: ndgnam@ctu.edu.vn; hvtminh@ctu.edu.vn; lvthinh@ctu.edu.vn; nvcngan@ctu.edu.vn

N. H. Trung
Institute of Climate Change, Can Tho University, Can Tho City, Vietnam
e-mail: nhtrung@ctu.edu.vn

S. Niculescu (ed.), *European Spatial Data for Coastal and Marine Remote Sensing*, https://doi.org/10.1007/978-3-031-16213-8_10

reduction efficiency of the current measures in the long term. It is necessary to monitor continuously and with different wind seasons to have a more accurate assessment of wave reduction efficiency.

Keywords Beach volume change · Shoreline erosion/accretion · Vinh Chau Town · Soc Trang Province · Wave height reduction

1 Introduction

The Mekong River is one of the longest rivers in the world, with almost 3000 km from its source in Tibet to the sea. The Mekong Delta begins at Phnom Penh in Cambodia, where the Mekong River meets the Tonle Sap River and the Bassac River tributaries. In Vietnam's territorial waters, the Mekong River divides into six main tributaries. The Bassac River divides into three main tributaries to form the nine dragons of the outer Vietnam Mekong Delta (VMD). The VMD region produces a considerable amount of rice and fish products and is the home of nearly 20% of the national population.

Being formed mainly on the geological setting of marine, river, and marsh sediments, combined with anthropogenic activities such as dam construction and sand mining [1, 5], the VMD has faced continuous natural/unnatural erosion impacts on both the riverbank and shoreline [2, 19]. Indeed, the total number of erosions on the VMD river and canal network is up to 665 points, with a total erosion length of 1048 km [12] that takes into account 1.15% of a total length of 91,000 km canal network [18]. Particularly in Soc Trang Province, a previous study by Pham [22] recorded 11.5 km coastline deposited among the eastern coordinates of 610,000–630,000 m (WGS-84) from the year 1965 to 2008. In some cases, the erosion features were associated with events that have caused the loss of human life, loss of productive aqua-agricultural land, and damage to infrastructure facilities.

For shoreline erosion in the VMD, there are several related studies conducted by national and international scholars. The earliest study on erosion in the VMD was undertaken using field survey—an expensive method—in November 2003 [5]. The study recognizes the large scale of erosion problems in the Mekong Delta, which are associated with many social and economic implications and consequences. In improving the understanding of occurrence and proposed solutions for river networks in the VMD, Hung [13] combined remote sensing (RS) and Geographical Information Systems (GIS) with mathematical models to quantify and predict erosion and sedimentation dynamics. Anthony et al. [2] applied high-resolution SPOT 5 satellite images to study shoreline erosion and land loss between 2003 and 2012, and they recorded that erosion affects over 50% of the once strongly advancing, over 600 km long delta shoreline. Another study using Landsat satellite images for the period of 1950 to 2014 showed progressively increasing of delta erosion, especially along the muddy East Sea coast, whereas the river mouth areas have shown a fluctuating trend of sediment supply [4]. Lately, Thuy et al. [27] recorded the largely changed of accretion and erosion processes in time and space of the Bassac and Mekong Rivers using MIKE 21 simulations as these processes are a combination of hydrological and sedimentation alterations due to upstream damming, sand mining, climate change, and sea level rise.

Besides that, wave reduction efficiencies of different structures were achieved by both site survey and measurement and modeling. Recently, Ty et al. [21] used the method of map overlaying in combination with field survey to assess the changes of coastal line and to evaluate the reduction of wave height through the Busadco breakwater at the eastern coastline of the Ca Mau Province for the period of 2008–2018. The results show that coastal erosion has been occurring very fast (on average about 50 m/year), and the Busadco structure reduces wave height of 1/10 H_{max}, 1/3 H_{max}, and Hmean at about 63%, 64%, and 67%, respectively. Loi et al. [15] used the AWH-USB to measure the wave pressure and then calculated the wave height through the mangrove forest at Tran De District, Soc Trang Province. They recorded that the coastal mangrove forest could reduce the tidal wave during both the high and low tide periods; the denser the forest structure, the greater the ratio of wave reducing and the lower the coefficient of wave reduction. Mai et al. [16] applied non-hydrostatic model Simulating Waves till Shore (SWASH) to assess wave reduction efficiency of bamboo fencing in Nha Mat Ward, Bac Lieu City. The results showed that a fence with 50% porosity reduced wave height by 11–72%, while a fence with 90% porosity reduced 29% wave height. An earlier study by Bao [3] defined minimum mangrove band width for coastal protection from waves in the Can Gio District of Ho Chi Minh City. The wave height was measured manually. He stated that wave height reduction depends on initial wave height, cross-shore distances, and mangrove forest structures.

As we can see, the mentioned studies solely focused to assess coastline erosion status or to evaluate the wave reduction efficiency of various structures but did not combine these topics. In addition, for administrative management purposes, local authorities need to quantify beach volume changes, but this information does not exist yet. According to Hanson and Kraus [10], to predict the beach changes over time, several models are available, ranging from detailed, micro-process-based two-dimensional and three-dimensional models to more engineering-office-oriented one-dimensional shoreline response (1-line) models and beach profile change models. Among the 1-line models, the GENESIS model has been applied more widely than any other model, exceeding installation at more than 1000 sites worldwide [9].

This study applied Landsat and Google satellite images from 2005 to 2022 to monitor the coastline erosion and to quantify changes of beach volume in Vinh Chau Town, Soc Trang Province. In addition, the study aimed to evaluate the shoreline protection potential of some current wave control structures at this study site.

2 Material and Methods

2.1 Study Area

The study was carried out in the Soc Trang Province, located on the eastern coast of the VMD region with 72 km of coastline (Fig. 1). Affected by a dynamic process of accretion and erosion created by the flow regime of the Mekong River and its sediment load, the tidal regime of the Vietnamese East Sea, and the prevailing monsoon winds, some parts of the coastline in Vinh Chau Town have experienced loss of valuable agricultural land due to natural and unnatural accelerated erosion processes [28], while in other parts of the region, accretion has occurred [14]. The mangrove forest belt that extends over 1900 ha of Soc Trang coastal zone has contributed to coastal protection between 1990 and 2000, but then, the forest disappeared due to rampant cutting leading to severe coastline erosion [23]. To reduce the negative

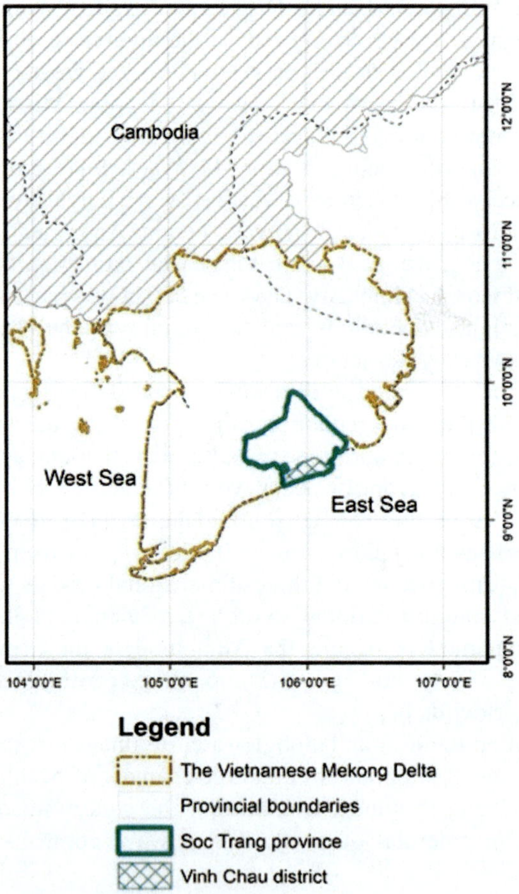

Fig. 1 Study area in Vinh Chau Town, Soc Trang Province, Vietnam, relative to the Vietnamese Mekong Delta region

impacts of the waves on people and farmland and shoreline recession on the Vinh Chau coastline, since 1995, a 55.89 km sea dike was built behind the mangrove forest belt as a second defense to protect the coastline [6].

Together with other provinces located in the coastal zone of the VMD region, the Soc Trang Province has been facing a rapidly growing shrimp production, especially in 2015–2019 with a yearly average increase of 9% of white leg shrimp and 1.2% black tiger shrimp installations [8]. The shrimp farming area extended through conversion of both coastal crop land [26] and mangrove forest [22]. As a result, the protection function of mangrove forest belt along the shoreline was reduced. Since 1995, an increase in the number and spatial extent of erosion in the vicinity of the sea dike and mangrove forest belts in coastal zone of Soc Trang Province has been recorded [23].

In response to the acceleration of coastal erosion, the local authority constructed a 1400 m concrete dike (breakwater) in the most eroded part of the coastline. The breakwater was built with a distance of 150 m in front of the sea dike for wave energy reduction testing purpose. The construction works started in June 2019 and finished on December of the same year.

2.2 *Data Collection*

2.2.1 Landsat Image Analysis of Shoreline Position

To quantify the variation in erosion features along the shoreline at Vinh Chau Town, Soc Trang Province, Landsat images from the years 2005 to 2022 were collected, pre-processed and analyzed. In this study, Landsat 5 Thematic Mapper (TM) Collection 2 Level 1 and Landsat 8 Operational Land Imager (OLI) and Thermal Infrared Sensor (TIRS) Collection 1 Level 1 were downloaded from US Geological Survey (USGS) server (https://earthexplorer.usgs.gov/). Details of the Landsat images are presented in Table 1.

Images downloaded from GoogleTM Earth Pro (http://www.google.com/earth/version/#earth-pro) were used in this study to compare with the shoreline variations obtained from the Landsat images. The GoogleTM Earth images have a better spatial resolution (1 m spatial resolution), but a smaller spatial extent each.

Table 1 Landsat images used to assess the variation in erosion along the shoreline and river mouths of the VMD

Year	Path-row	Satellite sensor	Spatial resolution (m)
2005	125–53/54	Landsat 5 TM	30
2009	125–53/54	Landsat 5 TM	30
2015	125–53/54	Landsat 8 OLI/TIRS	30
2020	125–53/54	Landsat 8 OLI/TIRS	30
2022	125–53/54	Landsat 8 OLI/TIRS	30

2.2.2 GoogleTM Earth Images

GoogleTM Earth images from the years 2006 to 2020 were used to digitize the specific beach part of the Vinh Chau shoreline (Table 2).

2.3 *Data Pre-processing*

2.3.1 Landsat Image Analysis of Shoreline Position

The steps undertaken for the image pre-processing are presented in Fig. 2.

First, tiles of the Landsat images of the same year were mosaicked without atmospheric correction. Next, the images were stacked in ENVI 5.3 (https://www.l3harrisgeospatial.com/Software-Techology/ENVI) and the Rotate/Flip function was used to clip the region of interest (ROI), which contains only the 72 km long shoreline along the Vinh Chau Town. Subsequently, the stacked images were imported to ArcMap 10.8 software (https://arcgis_desktop.en.downloadastro.com/) for shoreline extraction. From these, the shoreline positions were used to determine the variation in the location of the shoreline and calculate possible erosion and accretion zones and rates of the spatial extent of the sandy beach using the digital shoreline analysis system (DSAS) according to the guideline of Himmelstoss et al. [11]. In addition, the longshore sediment transport rate (LSTR) was calculated using the theory of the 1-line model [10].

2.3.2 GoogleTM Earth Image Analysis

As the GoogleTM Earth images are not in the same geometric frame, they were georectified using the Registration function of the Map tool in ENVI 5.3. Then, the corrected images were used to extract the shoreline positions in ArcMap 10.8 and calculate the shoreline change rate in DSAS as mentioned by Dolan et al. [7]. The method to analyze GoogleTM Earth image in this study was taken from the previous study of Malarvizhi et al. [17].

Table 2 Details of the GoogleTM Earth images in Universal Transverse Mercator (UTM)

Acquisition time	Sources	Resolution (m)	Projection
04/12/2006	Maxar Technologies	1.0	UTM
08/04/2014	French Space Agency (CNES)/Airbus	1.0	UTM
09/06/2017	CNES/Airbus	1.0	UTM
09/03/2018	CNES/Airbus	1.0	UTM
15/02/2019	Maxar Technologies	1.0	UTM
14/11/2020	Maxar Technologies	1.0	UTM

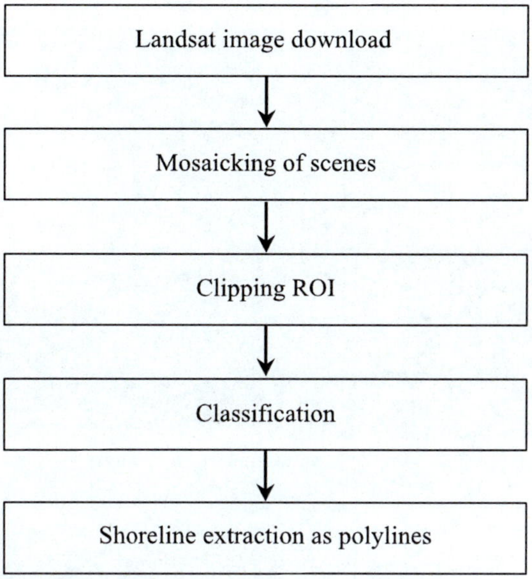

Fig. 2 Overview of Landsat image processing and analysis framework

2.3.3 Wave Measurements

Equipment: To measure wave pressures, the INFINITY AWH-USB data logger (https://www.jfe-advantage.co.jp/eng/assets/img/products/ocean-infinity/INFINITY-WH(E)_2016611.pdf) was installed in three sections of the shoreline, namely, A, B, and C. The sensors were attached to an iron pole and set submerged in the water body during flood tide to ensure the water layer was at least 20 cm above the sensors. The AWH-USB is provided with a storage memory capacity that allows for 0.1 s sampling intervals and wave height observations from short- to long-period waves.

Measurements of wave pressure were done at several locations. Wave pressures at section A and section B were measured on May 31, 2019, and June 2, 2019, respectively, to compare the wave reduction effectiveness of the mangrove forest belt and the sea dike. Wave pressure in section C was measured on January 15, 2020, to assess the wave height reduction through the breakwater. The installation of sensors and measurements varied across the three different sections (Fig. 3):

- Section A: The sea dike. Three sensors were installed at points 1, 2, and 3 with a 50 m distance between these three points and the shore. The water pressure was recorded at each point, and then, wave heights were calculated.
- Section B: The mangrove forest belt. Three sensors were installed (without, thin, and dense mangrove belt, respectively, at points 3, 2, and 1). The distance between the three points and the shore is about 50 m. The water pressure was recorded at each point, and then, wave heights were calculated.
- Section C: The concrete breakwater. Only two sensors were installed at a distance of 15 m in front (point 2) and behind (point 1) the breakwater. Water pressure was recorded, and then, wave heights were calculated.

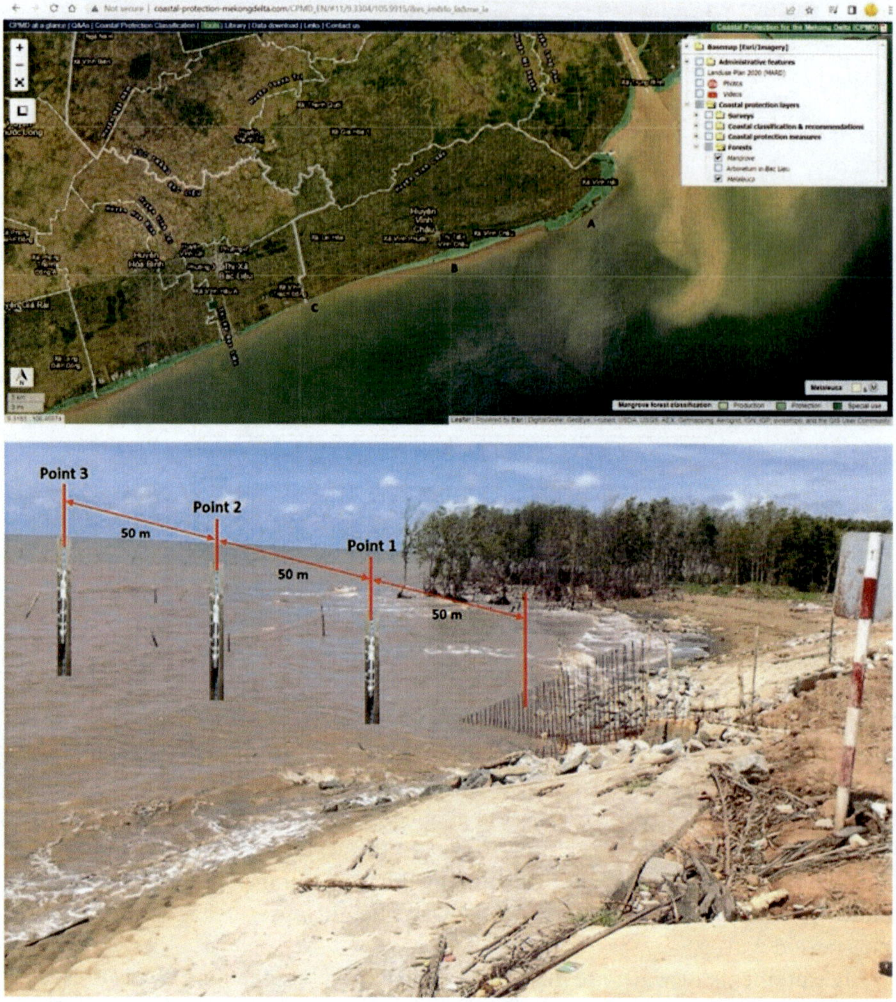

Fig. 3 Study sections and positions of wave measure equipment installation

Fig. 3 (continued)

2.4 Data Analysis

2.4.1 Change in Beach Volume

Understanding the sediment transport process along the coast is essential for better management and sustainability of this coastal zone [24]. For this reason, the change in beach volume where most erosion occurred (recorded from Landsat images) was evaluated based on GoogleTM Earth images covering 1.5 km of shoreline at the Vinh Chau coast. The shoreline profile was divided into nine cross sections (Fig. 4) to calculate the shoreline change rate indicated as a (m/year). The shoreline change rates in each cross section were calculated based on the temporal variation of the shoreline positions using the linear regression rate (LRR) method in DSAS.

The beach volume change was calculated by applying the theory of the 1-line model [9], in which the beach profile was assumed to move uniformly, as illustrated in Fig. 6. In this context, DB is the height of the berm (landward limit of the beach berm) and DC is the depth of closure (seaward limit of the active beach profile).

Based on the idea of 1-line model, a contour line can be used to adequately represent the beach plan shape for the calculation of the beach volume change, assuming the beach is sandy, not rocky, or with pebbles. This approach provides a simple link between the shoreline change and the spatial variation of the longshore sediment transport rate as the equation:

$$\Delta V = \Delta x \times \Delta y \times D \tag{1}$$

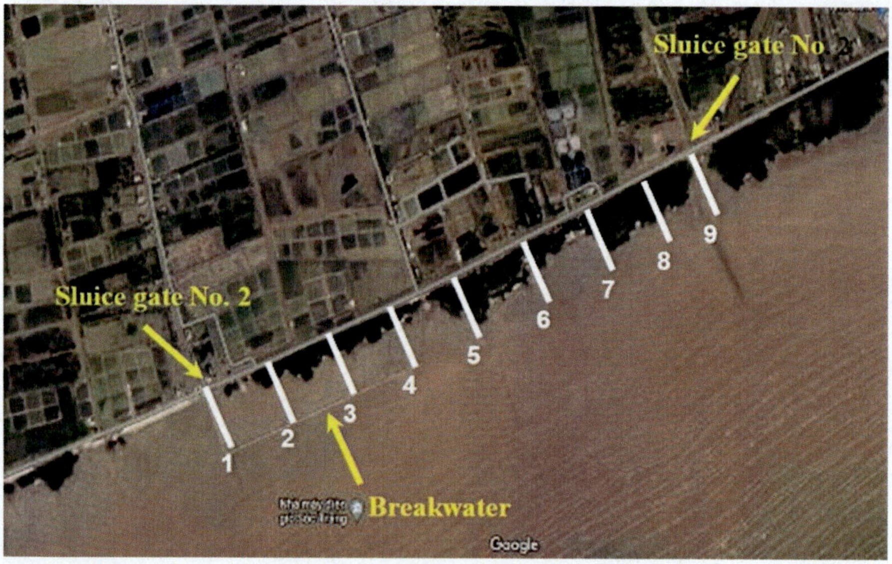

Fig. 4 The specific location to calculate the longshore sediment transport rate

where Δx is the distance along the coast (m), Δy is the distance offshore (m), $D = D_B + D_C$ (m), and ΔV is the change in beach volume (m^3/year).

2.4.2 Wave Height Calculations

Wave reduction assessment: The wave reduction efficiency was evaluated according to the national standard TCVN 12261:2018 hydraulic structures, coastal protection structures, design requirements for groin and detached breakwater [20].

$$\varepsilon = \left(1 - \frac{H_t}{H_s}\right) \times 100\% \tag{2}$$

where Ht is the wave height at point 3 and H_s is the wave height at points 2 and 1.

Data analysis: All recorded data were compiled and analyzed to calculate the wave values of $H_{\max}$, Hmean, 1/10 $H_{\max}$, and 1/3 $H_{\max}$, which represent the maximum, the mean, and the average of 10% and 33.3%, the highest wave height, respectively.

3 Results and Discussion

3.1 Results from Remote Sensing Images

3.1.1 Shoreline Changes from Landsat Images

The shoreline changes from the Landsat images between 2005 and 2022 are presented in Figs. 5 and 6. The analysis of shoreline position shows that significant erosion occurred about 8 km (Rectangular 1) from the border between the Soc Trang Province and Bac Lieu Province (from the eastern coordinates WGS-84 of 588,000 m to 596,000 m). The erosion rate varies from −16.9 to −3.0 m/year.

Another part of the shore showed predominant sediment accretion along the middle part of the shore from the eastern coordinates WGS-84 of 600,000 m to 620,000 m (Rectangular 2). The accretion rates vary from 45 to 80 m/year in the period 2005–2014, between 25 and 52 m/year in the period 2014–2022. Currently, the floodplain width at this part is about 600–810 m, creating an excellent zone to plant mangrove forest for coastline protection and maintenance purposes.

3.1.2 Shoreline Changed from Google Images

The results of the change in the shoreline from the years 2006 to 2020 using GoogleTM Earth images are presented in Fig. 7.

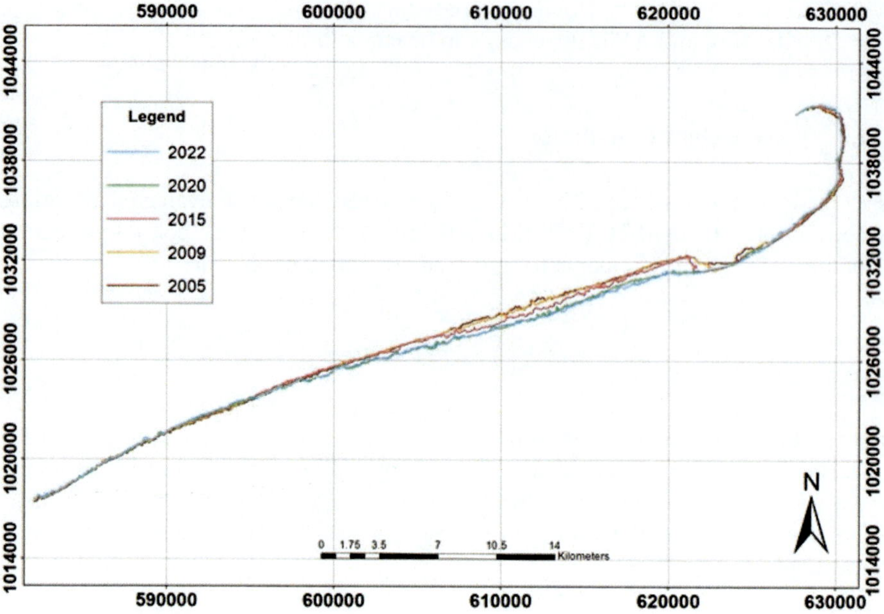

Fig. 5 Shoreline positions from 2005 to 2022

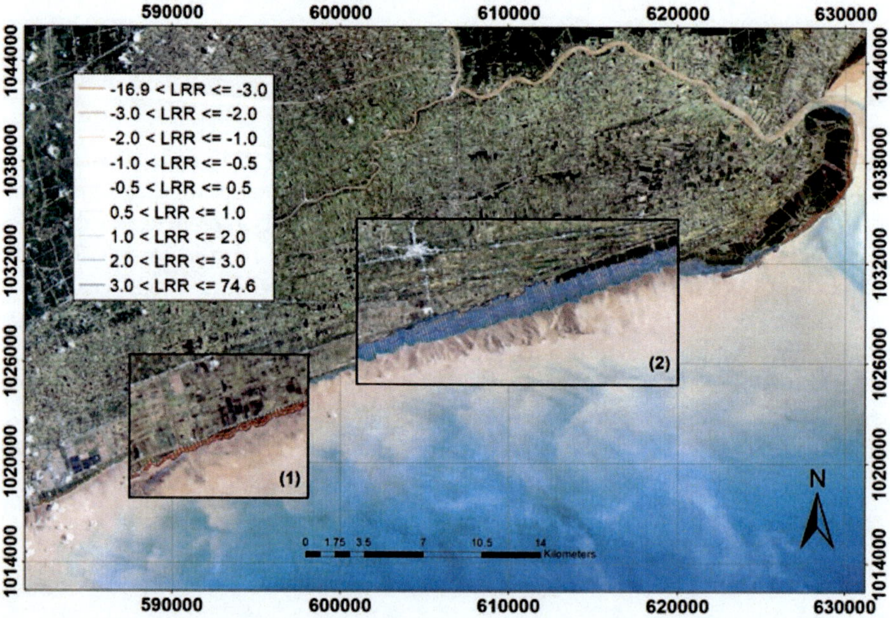

Fig. 6 The shoreline changes from 2005 to 2022

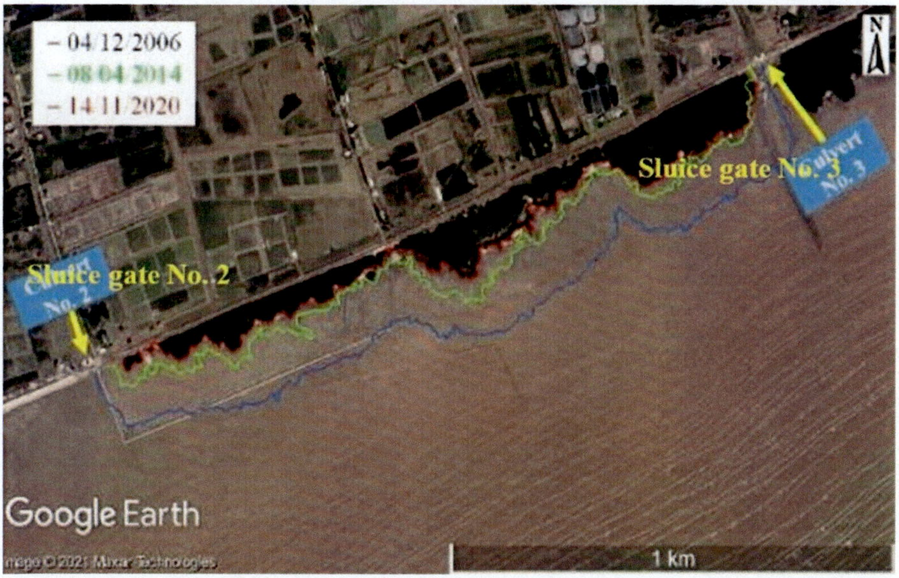

Fig. 7 Shoreline change from 2006 to 2020

It should be noted that only the shoreline positions for the years 2006, 2014, and 2020 are presented for the legible purpose. As can be seen in the figure, the shoreline has significantly retreated from the years 2006 to 2014 at distances of 70–140 m. During the period from 2014 to 2020, the shoreline retreated at smaller distances of 10–50 m.

3.2 Changes in Beach Volume

In Table 3, the shoreline change rate derived from GoogleTM Earth images varies from −11.68 to −7.95 m/year. These results are in line with the results obtained from Landsat images (−16.9 to −3.0 m/year).

The coefficient of determination R^2 values ranged from 0.8 to 0.98, which shows the close relationship between the offshore distance in each cross section by year (y) and year of study (t). Figure 8 presents the very high close relationship of y and t of the cross-section 1-1.

The change in beach volume was calculated using the shoreline change rate presented in Table 4 and Fig. 9. Negative values of the change in beach volume indicate that erosion occurred along the beach. The changes in beach volume ranged from 3000 to 7000 m^3/year.

Based on the changes in beach volume in this study, the LSTR value can be calculated to understand the erosion mechanism of the coastal zone. This can be done by taking the breakwater as a boundary where LSTR = 0 and integrating the changes in beach volume to find LSTR along the beach [25].

Table 3 Calculated shoreline change rate along the coastline

Cross section	Alongshore distance x (m)	Shoreline change rate a (m/year)	R^2
1–1	0	−7.95	0.98
2–2	200	−10.44	0.94
3–3	400	−8.58	0.98
4–4	600	−8.48	0.86
5–5	800	−8.16	0.97
6–6	1000	−11.68	0.93
7–7	1200	−9.44	0.90
8–8	1400	−8.76	0.89
9–9	1520	−10.75	0.80

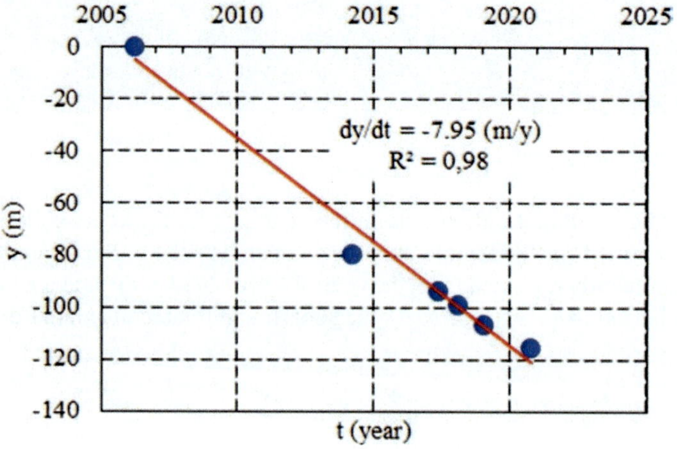

Fig. 8 Relationship between y and t of the 1–1 cross section

Table 4 Beach volume changes along the study coastline

Alongshore distance x (m)	Beach volume change ΔV (m^3/y)
100	−5518
300	−5707
500	−5119
700	−4992
900	−5953
1100	−6338
1300	−5460
1460	−3512

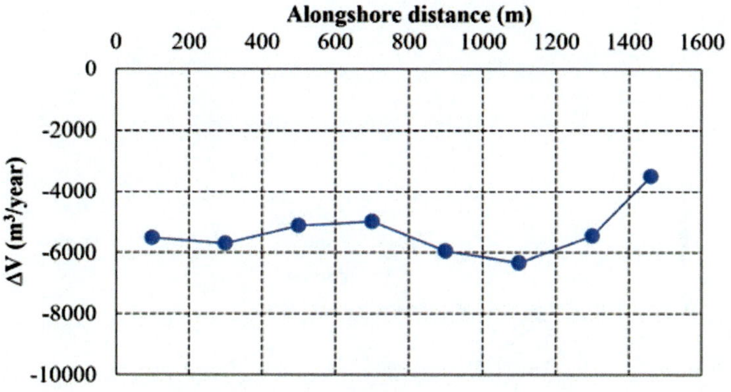

Fig. 9 Erosion/accretion rates at studied cross sections

3.3 Results on Wave Height and Wave Reduction

3.3.1 Wave Characteristics

The wave characteristics in sections A, B, and C are shown in Fig. 10. During the 4 hours of continuous recording of the water pressure (and thus calculated wave height), the wave height at different points of the three sections has changed. The recorded water height reflects the tidal regime of the East Coast (semi-diurnal tide). At the mangrove forest belt and breakwater sections (sections B and C), the recorded water level (wave height) further offshore is greater than the wave height near the coastline, showing that the wave height after passing through the mangrove forest belt or the breakwater has decreased significantly.

- At the sea dike (section A), the number of waves with height less than 10 cm accounts for the majority, and it decreases gradually from point 3 to point 2 and point 1. Wave height decreases in this section because of the slope (although very gentle), creating bottom friction in the direction of the wave to the shoreline, thus reducing the wave energy. At point 3, the wave height greater than 10 cm accounts for about 10%, and the wave height of 15 cm is about 5%. When the wave enters point 1, the wave height of more than 10 cm is about 10%.
- At the mangrove forest belt (section B), it can be seen that the number of waves with a height greater than 10 cm decreased significantly from point 3 to point 1. At point 3, a wave height greater than 10 cm accounts for approximately 10%; when a wave enters point 1, the wave height greater than 10 cm almost does not exist due to the wave reduction effect of the mangrove forest belt.
- At the breakwater (section C), there was a significant difference between wave height in point 1 and point 2. In front of the breakwater (point 2), it shows that the wave height ranging from 160 to 190 cm represents the majority. The wave heights are of 40–50 cm after passing through the breakwater (point 1). It shows that the wave height is significantly different between point 2 and point 1.

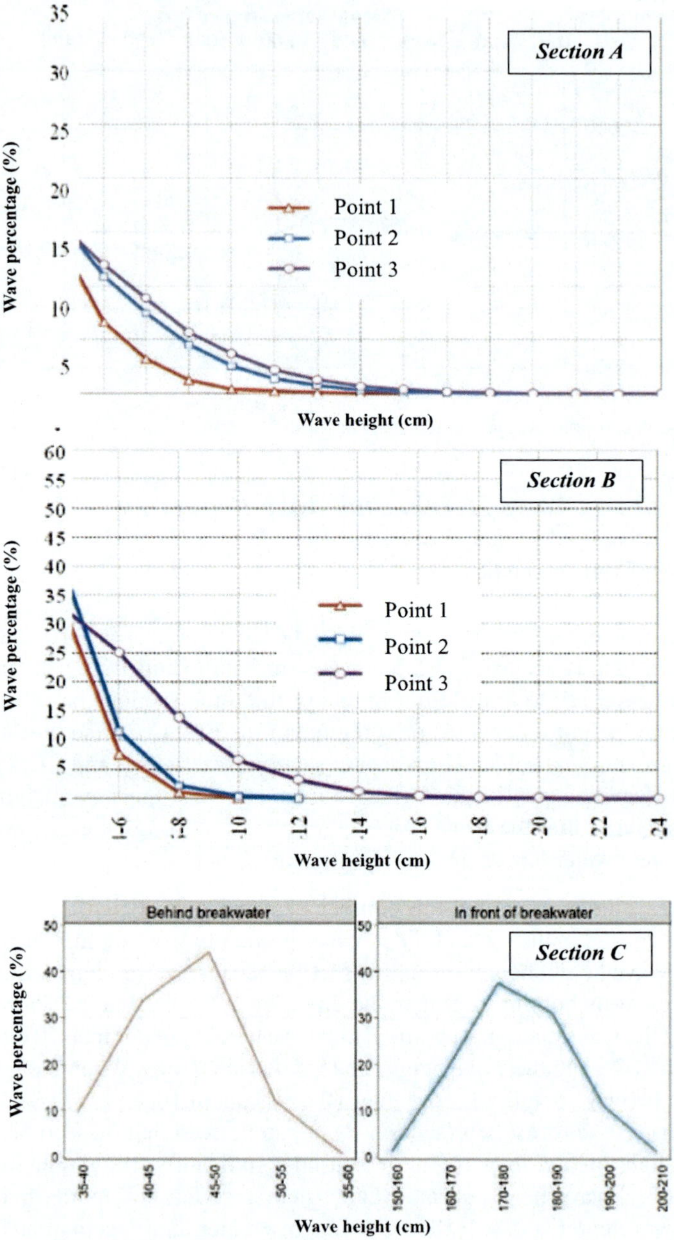

Fig. 10 Wave characteristics at all study sections

There was a difference in the wave height between the wave recorded in sections A, B, and C. The highest waves were measured in section C (Figs. 10 and 11), which occurred in the middle of January of the year. Indeed, November to April is a dry season in the VMD region, when the northeast monsoon wind is dominant with maximum wind speeds of 5–8 m/s. During this period, the wind intensity is the highest, causing high-intensity waves to hit the coastline at 45–50°. Therefore, the wave energy reduction was calculated in percentage values.

3.3.2 Wave Height Reduction

Wave height reduction in sections A, B and C is shown in Fig. 12. It shows the wave height reduction when entering the sea dike (section A) from point 3 to point 1 corresponding to H_{max} is about 36.8%, 1/10 H_{max} is about 34.4%, and 1/3 H_{max} is about 26.4% of actual measured data.

In the mangrove forest belt (section B), the wave reduction efficiency of the forest is relatively high from point 3 to point 1, matching H_{max} about 62.3% and 1/10 H_{max} about 55.3% and reaching about 54% to 1/3 H_{max} of actual measured data. Indeed, the wave height decreases by more than 50% when passing through the mangrove forest belt.

Wave height reducing efficiencies of the breakwater were the highest within the three measures. It is evaluated with the mean wave heights of Hmean, 1/3 H_{max}, and 1/10 H_{max}, and it can be seen that the breakwater wave reduction efficiencies are not significantly different. They are 72.18, 73.16, and 72.47%, respectively.

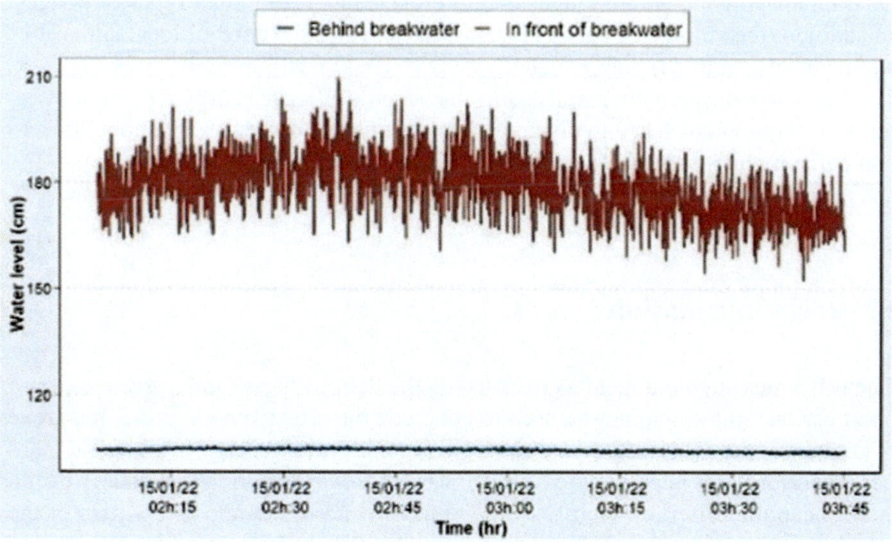

Fig. 11 Recorded water levels at section C

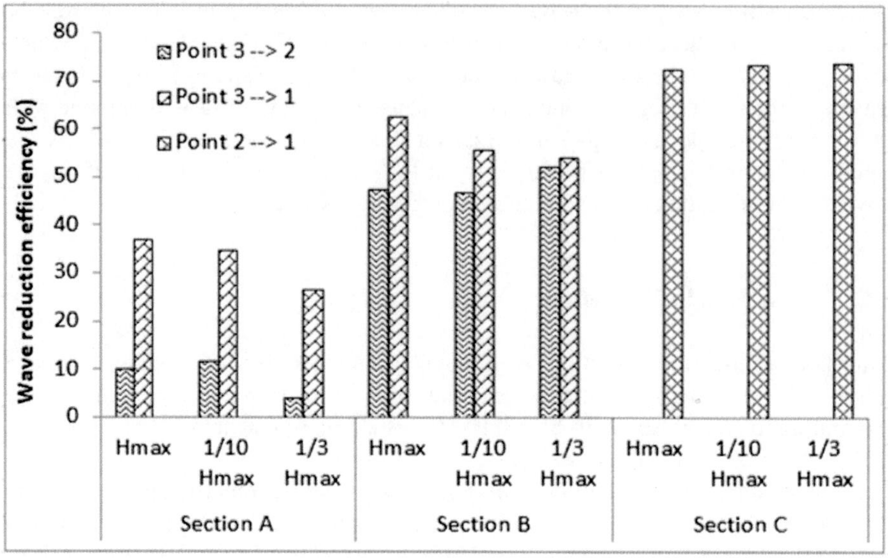

Fig. 12 Wave height reduction in various sections

4 Conclusion

Landsat satellite images and the GoogleTM Earth images can be used to study coastal erosion at Vinh Chau Town, Soc Trang Province, of the VMD region. The results showed comparable ranges of shoreline change from both types of images. In addition, the volume of beach changes that satisfies the need of local authorities was also defined by the 1-line model.

The wave measurement results show the benefits of wave energy reduction structures. Moving about a 150 m distance, the wave height decreases by about 30%. In comparison, the mangrove forest belt decreases the wave height by more than 50%, and the constructed breakwater has a better cut wave height by up to more than 70%.

5 Recommendations

The wave measurement results are based on the data collected and measured over a short period, so they cannot be used to conclude the effectiveness of the mangrove forest belt or the constructed breakwater in reducing wave energy in a general manner. Therefore, it is necessary to have a longer monitoring period (constantly over a year to capture different wind seasons) to have a more accurate assessment of the effectiveness of wave energy reduction through coastal forest.

In addition, forest characteristics such as density, stem and root height, and root area significantly influence the efficiency of wave energy reduction. Therefore, further research can consider the above effects on the wave energy reduction efficiency.

Acknowledgments The authors thank Erasmus+ INOWASIA project for the fund support to present these study results to the EUCOMARE 2021 Conference.

References

1. Anh NN (2018) Why does the river erosion situation become more complicated in the Mekong delta? Vietnam J Sci Technol Eng 60(1):73–82
2. Anthony EJ, Brunier G, Besset M, Goichot M, Dussouillez P, Nguyen VL (2015) Linking rapid erosion of the Mekong River delta to human activities. Sci Rep 5:14745. https://doi.org/10.1038/srep14745
3. Bao TQ (2011) Effect of mangrove forest structures on wave attenuation in coastal Vietnam. Oceanologia 53(3):807–818
4. Besset M, Brunier G, Anthony EJ (2015) Recent morpho-dynamic evolution of the coastline of Mekong River Delta: towards an increased vulnerability. Geophysical Research Abstracts Vol. 17, EGU2015-5427-1, EGU General Assembly 2015, Vienna
5. DDMFC – Department of Dike Management and Flood Control (2004) Bank erosion in Mekong Delta and Red River in Vietnam – Mission report. Ministry of Agriculture and Rural Development
6. Division of Water Resources and Storm Flooding Protection (2015) Facing to the coastal erosion in Soc Trang: challenges and solutions. Soc Trang's Department of Agriculture and Rural Development (in Vietnamese)
7. Dolan R, Fenster MS, Holme SJ (1991) Temporal analysis of shoreline recession and accretion. J Coast Res 7(3):723–744
8. GSO – General Statistics Office of Viet Nam (2019) Statistical yearbook of 2020. Statistical Publishing House
9. Hanson H, Kraus NC (1989) GENESIS-generalised model for simulating shoreline change – volume 1: reference manual and user's guide. Technical report. Coastal Engineering Research Center, U.S. Army Corps of Engineers.
10. Hanson H, Kraus NC (2011) Long-term evolution of a long-term evolution model. In: Roberts TM, Rosati JD, Wang P (eds) Proceedings, Symposium to Honor Dr. Nicholas C. Kraus. Journal of Coastal Research, Special Issue, vol 59. West Palm Beach (Florida), ISSN 0749-0208, pp 118–129
11. Himmelstoss EA, Henderson RE, Kratzmann MG, Farris AS (2018) Digital shoreline analysis system (DSAS) version 5.0 user guide. USGS's open-file report 2018–1179. https://doi.org/10.3133/ofr20181179
12. Hoai HC, Bay NT, Khoi DN, Nga TNQ (2019) Analyzing the causes producing the rapidity of riverbank erosion in Mekong Delta. Vietnam J Hydrometeorol 7:42–50. (in Vietnamese)
13. Hung LM (2004) Predict erosion and sedimentation changes and propose solutions for river networks in the Mekong Delta – scientific report. Southern Institute of Water Resources Research (in Vietnamese)
14. Joffre O, Luu HT (2007) A baseline survey in the coastal zone of Soc Trang Province, livelihood assessment and stakeholder analysis. Deutsche Gesellschaft für Technische Zusammenarbeit GmbH
15. Loi LT, Nguyen LT, Duy NN, Tri VPD (2019) Evaluating the wave reducing ability of mangrove thickness in Tran De district, Soc Trang province. Can Tho University J Sci Special Issue Environ Climate Change 2:18–26. (in Vietnamese)

16. Mai T, Dao T, Ngo A, Mai C (2019) Porosity effects on wave transmission through a bamboo fence. Proceeding of the 10th International Conference on Asian and Pacific Coast (APAC 2019). Ha Noi, September 2019. pp 1413–1418
17. Malarvizhi K, Kumar SV, Porchelvan P (2016) Use of high-resolution Google Earth satellite imagery in land use map preparation for urban related applications. Procedia Tech 24:1835–1842
18. Manh NV, Dung NV, Hung NN, Merz B, Apel H (2014) Large-scale quantification of suspended sediment transport and deposition in the Mekong Delta. Hydrol Earth Syst Sci Discuss 11:4311–4363. https://doi.org/10.5194/hessd-11-4311-2014
19. MARD – Ministry of Agriculture and Rural Development (2017) Land subsides and erosion in the Mekong Delta: Current status, reasons, and solutions oriented. Presentation at the Conference "Sustainable and climate-resilient development of the Mekong Delta of Vietnam", Can Tho city, Sep 26 – 27th 2017 (in Vietnamese)
20. MoST – Ministry of Science and Technology (2018) TCVN 12261:2018 Hydraulic structures – Coastal protection structures – Design requirements for groin and detached breakwater
21. Nghia NV, Minh HVT, Luan TC, Ty TV (2019) Assessment of wave reduction through the Busadco breakwaters: a case study in the East and West Coasts of Ca Mau province. Vietnam J Construct 6:584–591. (in Vietnamese)
22. Pham TT (2011) Mangroves of Soc Trang 1965–2007. Deutsche Gesellschaft für Internationale Zusammenarbeit GmbH
23. Pham TT, Hoang T, Tran HM, Le TH, Schmitt K (2009) Toolbox for mangrove rehabilitation and management. Deutsche Gesellschaft für Technische Zusammenarbeit GmbH
24. Rodríguez EL, Dean RG (2009) A sediment budget analysis and management strategy for Fort Pierce Inlet, Florida. J Coast Res 25(4):870–883
25. Tanaka H, Duy DV, Viet NT (2017) Evaluation of longshore sediment transport rate along the Thu Bon River delta coastlines in Viet Nam. Proceedings of the 37th IAHR World Congress (Kuala Lumpur). pp 3249–3255
26. Thach KSR, Vo HT, Lee JY (2021) Technical efficiency and output losses in shrimp farming: a case in Mekong Delta, Vietnam. Aust Fish 6(4):59. https://doi.org/10.3390/fishes6040059
27. Thuy NTD, Khoi DN, Nhan DT, Ngan TNQ, Bay NT, Phung NK (2019) Modelling accretion and erosion processes in the Bassac and Mekong River of the Vietnamese Mekong delta. Proceeding of the 10th International Conference on Asian and Pacific Coast (APAC 2019). Ha Noi, September 2019. pp 1431–1437
28. Trung NH, Woillez MN, Thanh ND, Eslami S, Minderhoud P, Quan TA, Hue NTT, Kien TB, Quang TC, Linh VTP, Thanh VQ (2021) Chapter 7 The Mekong Delta in the face of increasing climatic and anthropogenic pressures. In: Espagne E et al (eds) Climate change in Vietnam: impacts and adaptation. A COP26 assessment report of the GEMMES Vietnam project. Agence Francaise de Dévelopment, Paris, pp 339–369

Automatic Detection of Hydrodynamical and Biological Indicators of the Shoreline Using a Convolutional Neural Network

Soumia Bengoufa, Simona Niculescu, Mustapha Kamel Mihoubi, Rabah Belkessa, and Katia Abbad

Abstract The launch of satellites equipped with sensors in the optical range of the electromagnetic spectrum has greatly facilitated the mapping and monitoring of coastal areas for risk prediction. Thus, the frequent updating of information for monitoring purposes is possible. It is, therefore, a modern alternative to traditional methods, namely, photogrammetry and in situ investigation. The objective of this work is to define an efficient and validated method for the detection and extraction of shoreline indicators. It is the first indication of validation for a satellite image classification approach, based on a deep learning algorithm, optimized and adapted to the extraction, a hydrodynamic and biological indicator of the shoreline. The convolutional neural network (CNN) architecture was designed and adapted in order to extract the target shoreline indicators. A Pleiades image of very high resolution was used, sliced into sub-regions, and analyzed by a convolution kernel of size 3*3. The classification results have revealed a very high accuracy of 92%. A validation process was undertaken by comparing the results to field surveys (reference) acquired on the same day as the satellite image acquisition. With a run-up (horizontal wave excursion) of 0.6 m, the confidence interval for the deep learning method

S. Bengoufa (✉)
University of Western Brittany, CNRS, LETG Brest UMR 6554 CNRS, Technopôle Brest-Iroise, Av. Dumont d' Urville, Plouzané-Brest, France

National School of Marine Sciences and Coastal Management (ENSSMAL), University Campus of Dely Ibrahim Bois des Cars, Algiers, Algeria

S. Niculescu
Laboratory LETG-Brest, UMR 6554 CNRS, University of Western Brittany, Plouzané, France
e-mail: simona.niculescu@univ-brest.fr

M. K. Mihoubi
National Higher Institute of Hydraulics, Blida, Algeria

R. Belkessa · K. Abbad
National Higher Institute of Marine and Coastal Sciences, Algiers, Algeria

S. Niculescu (ed.), *European Spatial Data for Coastal and Marine Remote Sensing*, https://doi.org/10.1007/978-3-031-16213-8_11

was estimated to be ± 0.42 m, which is quite small, revealing the good accuracy of the method tested. A large panel of users could reproduce these methods in an automatic and standard way, which should allow the updating of a possible database shared between involved parties in an efficient way.

Keywords Convolutional neural network (CNN) · Deep learning · Shoreline detection · High water level · Marine lichen

1 Introduction

The coastal fringe is a space characterized by its permanent mobility on the scale of geological and human time. Today, this coastal fringe is heavily exploited and has become a territory at risk due to marine submersion and erosion hazards, which are evident on a global scale. Luijendijk et al. [16] derive an updated global assessment of erosion phenomenon. It is estimated that from 1984 to 2016, approximately 24% of the world's sandy beaches were subjected to very high erosion. Cliff coasts, which can only form by recession, must also be taken into account when discussing erosion.

The spatiotemporal analysis of shoreline position is among the most validated approaches for the characterization of the erosion phenomenon [7]. This analysis can be effectively carried out through the use of multitemporal aerial photography data, Geographic Information System (GIS) tools, and satellite image processing [21]. Historically, detection of the shoreline has been carried out by direct approaches on the ground or indirectly using an iconographic data (aerial photographs, images, maps, etc.) [7]. Several studies have examined the potential of satellite imagery to detect and assess shoreline evolution. Optical imagery has been shown to offer more practical approaches to detect the position of shoreline at different spatial and temporal scales [2, 12], especially with the development of very high resolution (VHR) multispectral sensors.

Although the relevance of the processing results is directly related to the resolution and type of satellite data, the development of accurate techniques for the detection of the shoreline position is ongoing. The most common techniques for shoreline detection were either the ground survey or manual digitization, the latter still being in common use. However, shoreline position detection and delineation are a tedious task and highly subjective when using traditional ground survey methods or the manual approach. The alternative is to use more automatic image processing techniques, such as supervised and unsupervised image classification. These (semi) automatic techniques are needed to update and extract shoreline data and make coastal land cover mapping more accessible.

Recently, several machine and deep learning algorithms have been successfully adopted for remote sensing applications [19]. This process translates human cognition to machine intelligence in a more sophisticated way [10]. Bengoufa et al. [5, 6] have attempted to assess the robustness of machine learning algorithms, i.e., the support vector machine and random forest (RF) algorithms, for the automatic detection of sandy shorelines. However, there are several shoreline indicators, the choice of which depends on the context and the objectives of the study and the geomorphological type of the coastline.

The shoreline is supposed to represent a linear boundary between the marine and terrestrial domains, but this remains problematic due to the heterogeneity of the identification criteria (geomorphology, tidal conditions, vegetation, etc.). Indeed, it is not a theoretical definition of the shoreline, which can be generalized to all environments, which should be given, but a "functional" definition of the coastal fringe according to the type of coast and the monitoring objectives [17]. Shoreline detection and definition have been the subject of several studies [7, 8, 20]. Boak and Turner [7] identified 19 generic shorelines based on 45 surveyed indicators. Indeed, the effectiveness of automatic methods for the detection of several types of shoreline has not been sufficiently studied. Indeed, through a comparative approach, we attempt in this work to evaluate the contribution of multispectral images and deep learning algorithms for the delineation detection and extraction of two commonly used indicators of shoreline.

2 Methodological Approach

2.1 Shoreline Indicator

The detection and analysis of the shoreline position require, first of all, a definition and understanding of the concept of a "shoreline." In our case, the term "shoreline indicators" is used. In fact, the choice of a shoreline indicator should meet the main conditions for recognition, which are the following:

1. Easily and uniformly recognizable on black and white and color images.
2. Linearly continuous along the beach.
3. Variations in its position along the beach due to water level changes are minimal.

In order to choose adequate shoreline indicators, an analysis of the geomorphological type of coastline existing in our study area was performed. Our study area is located on the western coastline of Algeria, in the Mostaganem province, 400 km west of the capital Algiers. It boasts a 124 km long coastline, an interesting landscape, and geomorphological diversity (Fig. 1). The geomorphological analysis revealed the existence of wide sandy beaches and low cliffs and artificial coastline.

Indeed, two shoreline indicators have been chosen, a hydrodynamic indicator for sandy coasts and a biological indicator for low cliff coastlines. Furthermore, for the artificial coastline, the terrestrial edge was taken as the target shoreline indictor.

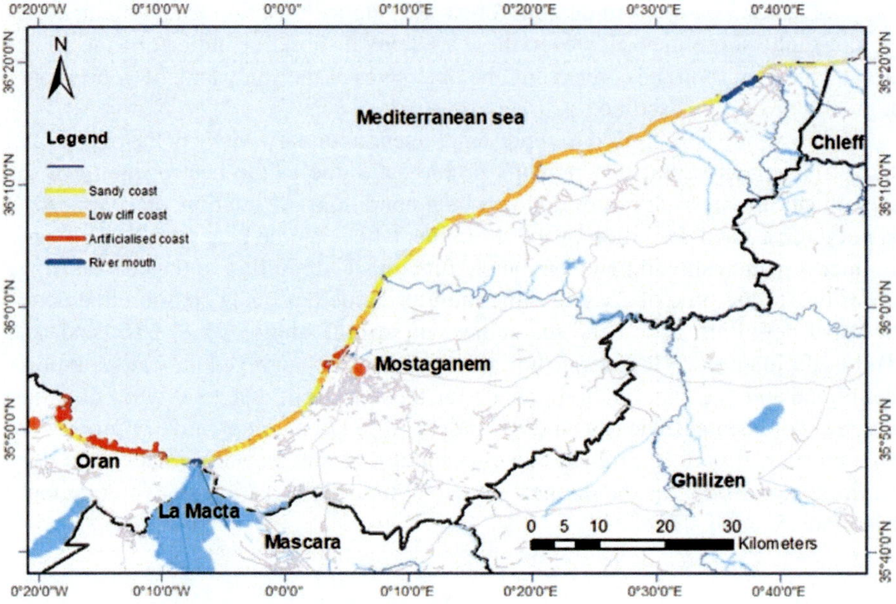

Fig. 1 Coastal geomorphological types in the study area (western Algeria)

Fig. 2 Shoreline indicators. (**a**) Hydrodynamical indicator for sandy beaches, high water level (yellow) (Bengoufa et al. 2021a). (**b**) Biological shoreline indicator for rocky cliffs, black lichen (yellow)

2.1.1 Hydrodynamical Indicator

The high water level (HWL) was found adequate as a hydrodynamic indicator of the shoreline for the sandy coast (Fig. 2). For a microtidal coast, HWL can be the same as the wet/dry sand limit [7], which can be detected on ground or satellite imagery. In the latter, the high reflectance of the sand results in high luminance. When submerged in sea water, the beach appears darker in the image. It thus has different grey values at infrared wavelengths, forming interpretative signals, which allows for effective detection on satellite images.

2.1.2 Biological Indicator

Furthermore, the supra-littoral level of the low cliffy coasts is characterized by the presence of black lichens. The literature refers to a species that could be the one detected in the field, *Verrucaria amphibia*, or *V. maura*, which appears as a black band on the rocky substrate (Fig. 2). Indeed, the upper limit of this band is considered a good biological/botanical indicator of the cliffy shoreline [8].

2.2 Data Set

In this study, we test the shoreline extraction performance of a deep learning model, i.e., convolutional neural network (CNN), on satellite images of the studied area. We attempt to assess the performance of an automatic technique of shoreline extraction by comparing it to in situ techniques. For that, very high resolution images from the Pleiades satellite were used (Table 1).

Moreover, in situ data were collected the same day as the Pleiades images acquisition and used to train, test, and validate the CNN model. The field trips were specially scheduled to coincide with the Pleiades image acquisition missions (1 July 2019) in order to assess the contribution and limitations of remote sensing in extracting fundamental information for monitoring coastal evolution. Thus, topographic surveys were carried out by a differential GPS (DGPS), whose measurements are made with centimetric precision.

The DGPS consists of a base GPS receiver, positioned on a point whose coordinates are precisely known (with a very high degree of accuracy) and another mobile GPS. The surveys were carried out with a type of real-time (instantaneous) coordinate measurement in real-time kinematic (RTK) mode.

For the shoreline indicator surveys, the point acquisition interval was one second in continuous topo (Fig. 3). The ground-truth data of the shoreline were compared to shorelines extracted by remote sensing methods, by calculating the

Table 1 Spectral and spatial resolutions of Pleiades images

	Pleiades	
Band	Wavelength (nm)	Resolution (m)
B1 (blue)	430–550	2
B2 (green)	500–620	2
B3 (red)	590–710	2
B4 (near infrared)	740–940	2
B5 (panchromatic)	480–830	0.5

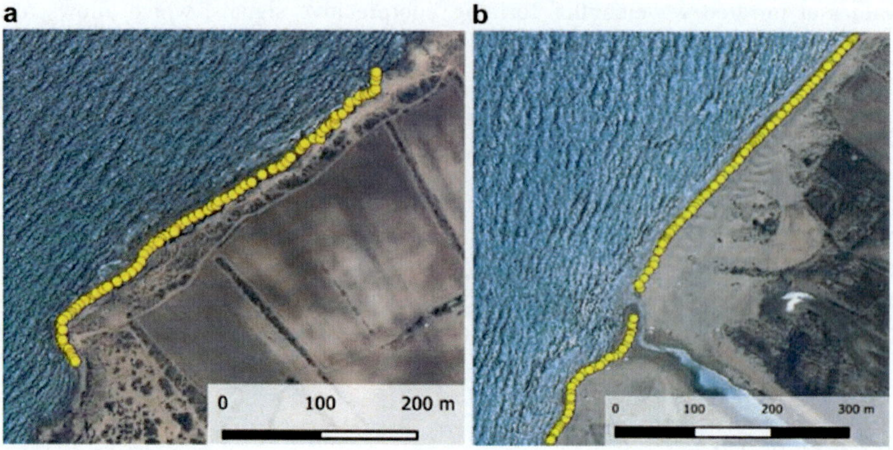

Fig. 3 Field surveys of shoreline indicators along different types of environments; (**a**) sandy coast, (**b**) low cliff coast

distances between the same indicators from each method (remote sensing and field survey).

2.3 *Convolutional Neural Network Architecture*

The convolutional neural network (CNN) [15] was used to process the Pleiades images. It is a deep learning algorithm adopted in a wide range of aspects of image processing [14]. The architecture of the CNN is a series of convolution and max-pooling layers. In this study, the CNN prediction model (Fig. 5) was designed and trained using eCognition software. A Pleiades image was used as an input, sliced into sub-regions or tiles, and analyzed by a convolution kernel of size 3*3.

The analysis of the image features by this kernel is a filtering operation with an association of weights to each pixel to produce a feature map. These represent very specific features used to determine classes for each shoreline indicator detection criteria: water, dry sand, and wet sand for HWL detection and water, black lichen, soil, and vegetation for biological indicator detection. In each class, each unit is connected to patches in the feature maps of the previous layer by shared weight matrices.

A data set of training sample points was used to create sample patches selected from the input image pixels (Fig. 4).

To choose an adequate sample size, we took into consideration the dimensions of the target classes (e.g., the lichen class for cliff coast, the wet sand class for sandy coast).

After a cross-validation process, the CNN architecture was fed with 4×4 sample size patches. Given this small size, there were two hidden layers (Fig. 5), with a

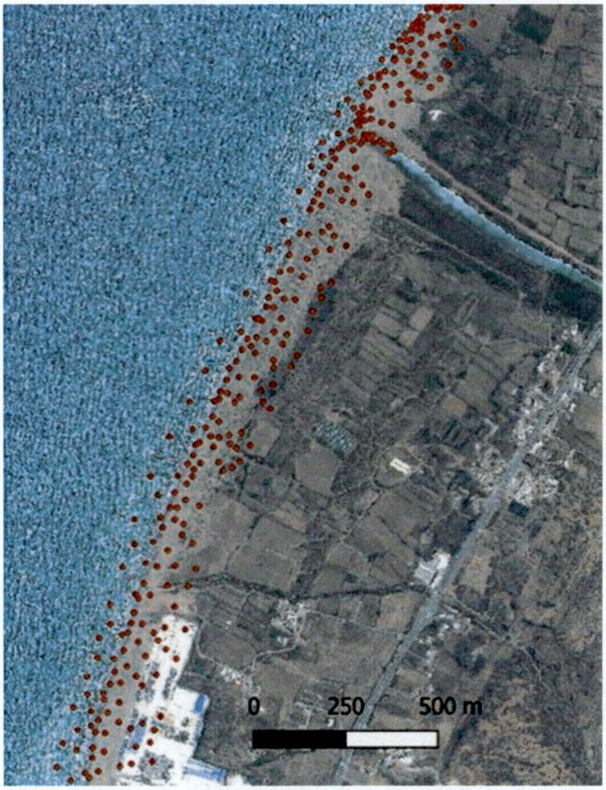

Fig. 4 Representation of the training sample points (in red)

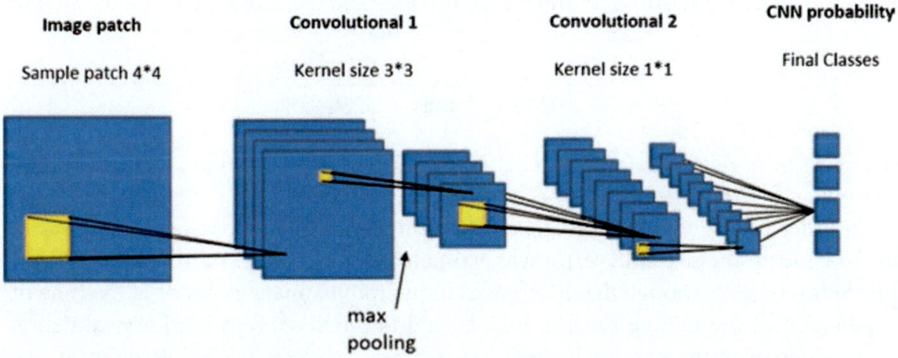

Fig. 5 Presentation of the architecture of the CNN model used (eCognition, 2020)

convolution kernel size of 3 × 3 and then 1 × 1. A max-pooling layer with a 2 × 2 kernel size was used after the first convolution layer. The role of max-pooling is to merge semantically similar features. It is a non-linear top-down sampling technique that uses the maximum value of each cluster in the previous layer to reduce the loss of information in the subsequent convolution layers.

In addition, a patch size of 50 and 5000 training steps resulted in good performance. A cross-validation process revealed that the optimal training rate is 0.0006. It should be noted that an inadequate learning rate can either increase the time of the learning process (at a low rate), which could lock the network into local minima, or decrease it (at a high rate), but the network might not reach the minima. Therefore, in both cases, it is possible to obtain incorrect weights.

After a cross-validation process, CNN architecture was designed for each shoreline indicator type. After a final classification, the results were converted to a polyline format in order to select and thus extract the target shoreline indicators.

2.4 Accuracy Assessment

In order to assess the accuracy of the CNN model in terms of shoreline indicator extraction, a comparison of its performance to another machine learning algorithm, the random forest algorithm, was performed. Thus, along with the analysis of the overall accuracy index, the shoreline indicators extracted from both methods were compared to the in situ reference data. That is to say, a calculation of the distance between the shorelines from the random forest and CNN models and the reference in situ shoreline was made, using the digital shoreline analysis system (DSAS) (Fig. 6).

Some uncertainties relating to the improvement of image location accuracy, DGPS accuracy, mean slope, and tide level may lead to a deviation of the extracted shoreline from the in situ shoreline. Therefore, an estimation of the confidence interval should be performed. The maximum total uncertainty is given by the square root of the sum of the squares of the values of the error sources:

$$\sqrt{\mathrm{RMS}^2 + \mathrm{Rmax} + \mathrm{GPS}^2} \quad (1)$$

where *RMS* is the georeferencing accuracy of the images, *Rmax* is the maximum run-up, and *GPS* is the accuracy of in situ data (±0.03 m).

In order to evaluate the influence of the improved image localization accuracy, the root mean square (RMS) error was estimated using ground control points (GCP). Furthermore, even though the differences in hydrodynamic conditions at the time of acquisition of the remote sensing images and the field surveys used for validation (5 hours apart in the case of Pleiades) theoretically have a negligible effect in the case of a microtidal coast (the tidal range is insignificant by 0.17 m according to Ali [1]), it makes sense to estimate the uncertainty and the influence of wave run-up on the horizontal position of the shoreline in order to validate the HWL extraction method. To this end, a hydrodynamic characterization was performed.

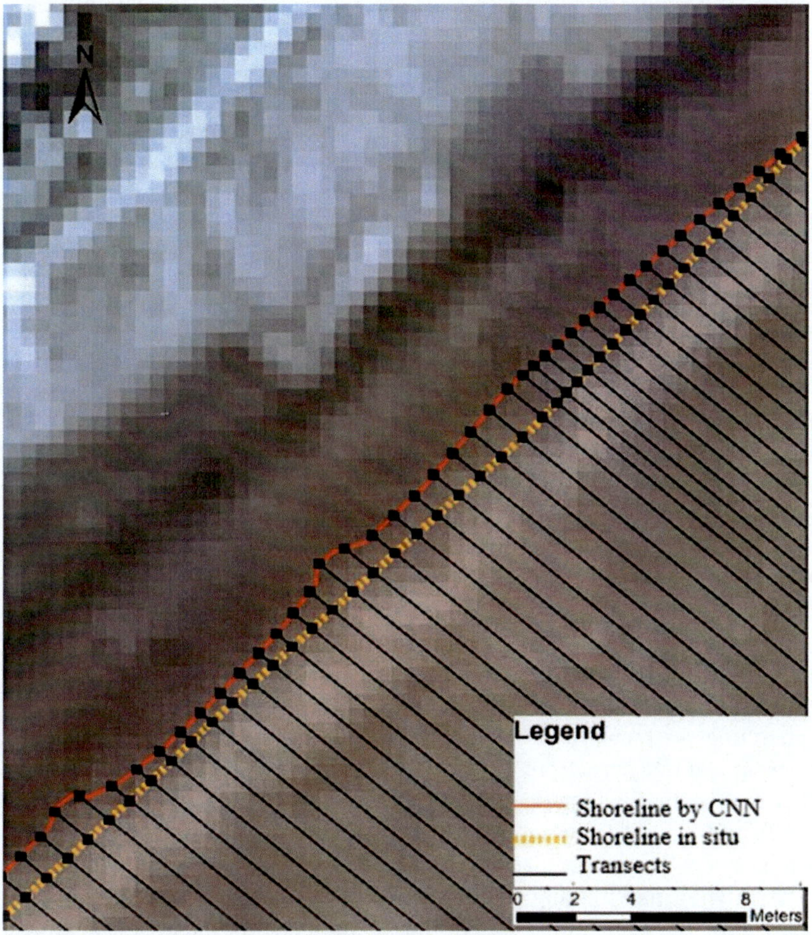

Fig. 6 Distance calculation (using DSAS Net Shoreline Movement-NSM) between the automatically extracted and the reference (in situ) shorelines

2.4.1 Hydrodynamic Characterization

After extracting the shoreline vectors for the sandy coastline (hydrodynamic indicator), we calculated the maximum run-up and the maximum horizontal wave excursion. Indeed, a numerical simulation of the hydrodynamical process was performed using the DHI MIKE 21 model in order to calculate the maximum run-up. The run-up is calculated as a function of the average slope of the beach ($\tan\beta$), the average height (H_{mo}), and the wavelength of the swell (H_{mo} and L_o) [13]. The use of these three variables has been simplified based on Iribarren's ξ_o number (or surf similarity parameter) [4]:

$$\xi_o = \tan \beta / \left(H_{mo} / L_o \right)^{1/2} \tag{2}$$

$$L_o = 1.561 \mathrm{ms}^{\dagger}.T_{m02\dagger} \tag{3}$$

where H_{mo} is the mean wave height, L_o is the offshore wavelength, and T_{m02} is the mean period of the swell.

The slope of the beach was calculated based on in situ topographic measurements from the top of the dune to about half a meter below the sea level. The intertidal beach slope was considered invariant in our study and was calculated from the average profile along the outer face of the berm.

The offshore wave data (H_{mo} and T_{m02}) were acquired by numerical simulation using the DHI MIKE 21 model. This type of simulation takes into account local wave generation due to wind. At the time of the DGPS survey of the shoreline, the offshore wave height was 0.2 m for a period of 8 s according to the hydrodynamic database developed for the Algerian coast by Amarouche et al. [3].

The horizontal wave excursion requires the conversion of the estimated run-up (vertical) elevation into an associated horizontal variation using the beach slope.

3 Results

The shorelines extracted as a result of the application of the CNN model using Pleiades optical images are presented in (Fig. 7).

This is a (semi-) automatic extraction method of the two shoreline indicators most effectively used to monitor coastal dynamics and characterize erosion phenomenon. Indeed, an image classification based on a deep learning model has been undertaken using, as a training data set, in situ samples collected on the same day as the acquisition of satellite images. The classes assigned for the image classification process were highlighted with an accuracy of 92%.

The shoreline indicators extracted from the Pleiades images (Fig. 7) were compared to the reference shoreline from the classical in situ methods. This comparison allows the validation of the deep learning models tested in this study.

Some uncertainties related to image location, GPS accuracy, and tide level may lead to a deviation of the extracted shoreline from the in situ shoreline. In order to quantify the influence of improved image georeferencing accuracy, the root mean square error (RMSE) was estimated using ground control points (GCP). The resulting RMSE was less than one-half of the pixel size (RMSE = 0.25 m). In addition, the GPS used for the in situ survey has centimetric accuracy (±0.03 m), comparable to other sources of uncertainty.

The main sources of error are greater for the hydrodynamic indicator due to its fluctuating nature. It was difficult to schedule an image acquisition on the same day as the in situ measurements without an interval of a few hours (five-hour interval). Considering the tidal effects, this interval could affect the accuracy of the validation procedure.

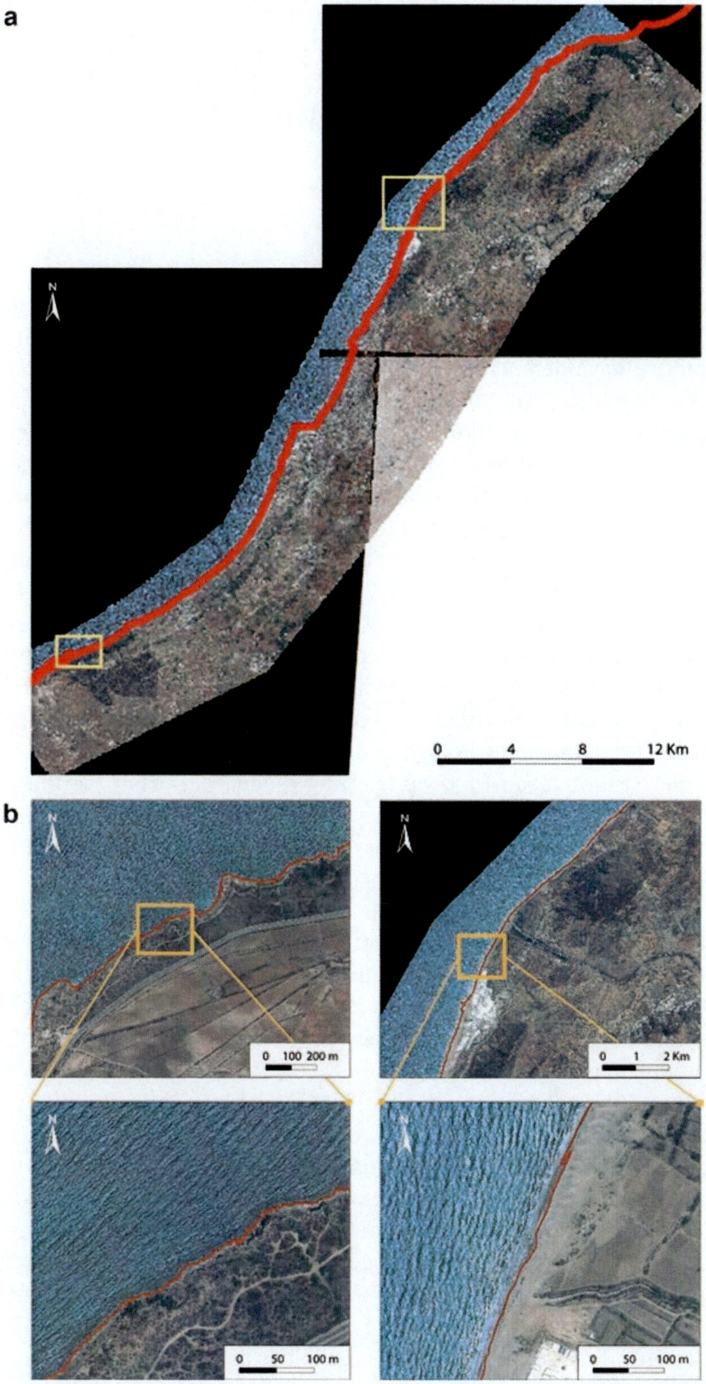

Fig. 7 Extracted shorelines by the CNN model. (**a**) Hydrodynamic indicator for sandy shorelines (HWL); (**b**) biological indicator for cliffy coasts (black lichen)

The tidal effect depends on the slope of the beach, which defines the maximum wave run-up. Considering average offshore wave heights of 0.75 m, for an average beach slope of 5°, we found that $\xi o = 0.73$ (Eq. 2), the run-up = 0.6, and thus, the total confidence interval ≈ 0.42 m calculated by the Eq. (1).

4 Discussion

The analysis of satellite data used to be the exclusive domain of researchers. Today, the increased availability and accessibility of satellite data, combined with powerful analytical tools, have democratized data innovation. This has opened up new possibilities for exploiting data and converting it into meaningful information. The increasing availability of satellite data has transformed the way we use remote sensing analysis to understand and monitor the coastal dynamic, to better prevent the coastal hazards and achieve the Sustainable Development Goals (2030 target).

The shoreline is a key information for erosion hazard characterization. It can be identified by several physical indicators detectable on remote sensing images (vegetation boundaries, water/sand boundary, cliff edge, and high water level [HWL]). For the assessment of shoreline changes, the sandy beach survey, and coastal erosion characterization, the HWL, which corresponds to the highest limit reached by seawater, was the most used physical indicator. This indicator has historically been detected using manual digitization, though recently several authors have attempted to detect and extract this indicator using a more automated technique. In this work, we tested the performance of a deep learning model, i.e., CNN for the automatic extraction of the HWL. It should be noted that the CNN model achieved a good performance for the detection in the classification process, with an accuracy index of about 92%.

Furthermore, as the rocky cliffs are more complex and heterogeneous, the biological indicator was used for these sections. The chosen indicator is somewhat difficult to detect because of its fit in a rather narrow band along the coast at the foot of the cliffs. It is a widely used indicator relating to the black lichen belt. This complexity turned out to have little effect, as the CNN model was able to detect this indicator with good accuracy (91.6%).

We evaluated the contribution of the CNN model by comparing it to a machine learning algorithm, random forest, and compared the results to a reference in situ shoreline. The performance of deep learning algorithms is superior to that of machine learning. Indeed, it has been found that 76% of the shoreline extracted by the CNN model is located within one meter (1 m) of the reference (in situ) shoreline, as compared to 63% for the RF model (Fig. 8). However, the understanding of these algorithms and their interpretation is generally limited at this time [9, 18]. Therefore, the major point to consider is the difficulties of interoperability when using these algorithms. It can be said that the better the learning of an algorithm, the more difficult its interpretation (Fig. 9).

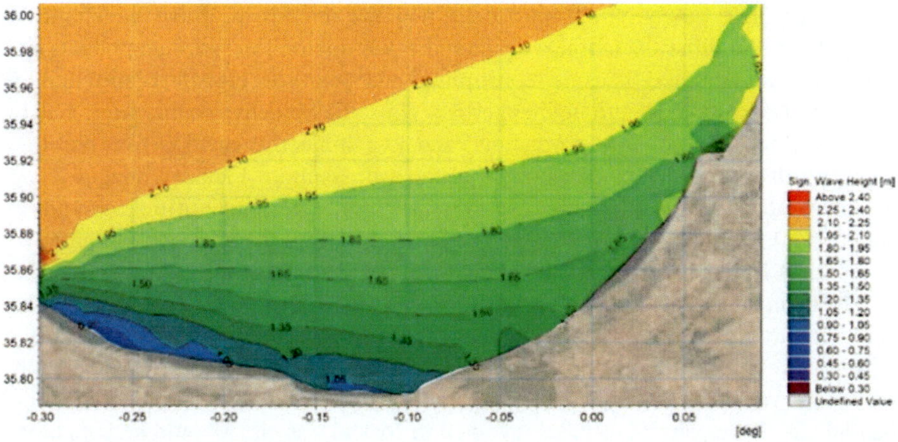

Fig. 8 Hydrodynamical simulation and wave heights calculated using the DHI MIKE 21 model

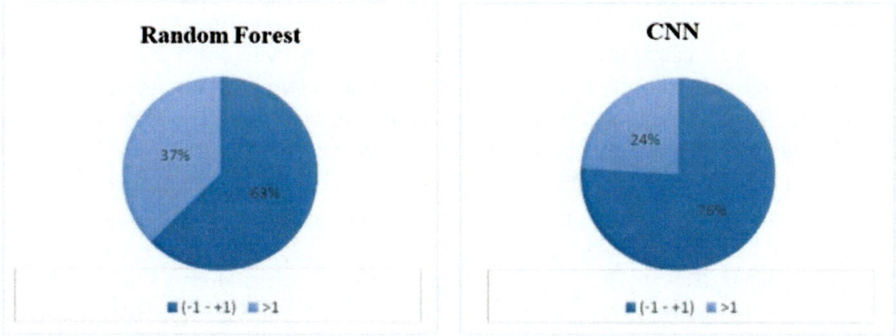

Fig. 9 Percentage of the extracted shoreline within 1 m of the reference shoreline (in situ)

Our methods were tested on a microtidal coast, with insignificant tidal range (0.17 m) [1]. In the case of shoreline detection for a macrotidal coast, this method would not be suitable without including an accurate estimation of the tidal range.

5 Conclusion

The present work is a contribution to the study of the coastal erosion risk through the detection of essential information, the shoreline indicator. We focus on methodological development based on the contribution of new remote sensing methods and highlight the progress that has been made in the use of high spatial resolution satellite data and artificial intelligence methods (deep learning).

For shoreline detection, in situ measurements remain the most widely used method for shoreline detection, considered a key factor for various applications. However, this method is time consuming and requires logistical and human resources with high cost and field expertise. Thus, the objective of this work was to determine an operational, reproducible, and less subjective method for shoreline detection. It was shown in this study that the processing of Pleiades images using the CNN model offered an immediate overview of the shoreline position, providing an alternative for coastal monitoring. The confidence interval has been estimated to be less than 1 pixel of the image by taking into consideration several sources of error.

The comparative study of two algorithms used for image processing revealed the need for further methodical development efforts in order to improve the reliability of remote shoreline monitoring and extraction. The classification based on the CNN model showed higher accuracy as compared to the random forest model and could be replicated by a variety of users involved in coastal issues. It should be noted that the results were obtained on a microtidal coast, with insignificant tidal conditions.

Coastal monitoring using new deep learning algorithms adapted to image processing could significantly optimize the time and costs of traditional methods and reduce the subjective component of manually digitizing shorelines from images. In addition, many research areas could benefit from improved shoreline extraction, such as coastal vulnerability assessments, erosion rate estimations, the development of numerical hydrodynamic models, environmental risk analysis, marine spatial planning, coastal management, and engineering. Furthermore, a graphical interface could then be developed for the model, which could also be integrated with the image processing software to allow interactive operations to improve the results.

Acknowledgments This study is supported by the ERASMUS+ Jean Monnet Chair, European Spatial Studies of Sea and Coastal Zones -599967-EPP-1-2018-1-FR-EPPJMO-CHAIR.

References

1. Ali R (2016) Traitement du signal marégraphique par analyse harmonique pour la détermination d'une référence altimétrique pour l'Algérie. BSG 19:2–10
2. Almonacid-Caballer J, Sánchez-García E, Pardo-Pascual JE, Balaguer-Beser AA, Palomar-Vázquez J (2016) Evaluation of annual mean shoreline position deduced from Landsat imagery as a mid-term coastal evolution indicator. Mar Geol 372:79–88. https://doi.org/10.1016/j.margeo.2015.12.015
3. Amarouche K, Akpınar A, Bachari NEI, Houma F (2020) Wave energy resource assessment along the Algerian coast based on 39-year wave hindcast. Renew Energy 153:840–860. https://doi.org/10.1016/j.renene.2020.02.040
4. Battjes JA (1974) Surf Similarity 1:26–26. https://doi.org/10.9753/icce.v14.26
5. Bengoufa S, Niculescu S, Mihoubi MK et al (2021) Machine learning and shoreline monitoring using optical satellite images: case study of the Mostaganem shoreline, Algeria. J Appl Remote Sensing 15. https://doi.org/10.1117/1.JRS.15.026509
6. Bengoufa S, Niculescu S, Mihoubi M et al (2021) Rocky Shoreline extraction using a deep learning model and object-based image analysis. Int Arch Photogrammetry

Remote Sensing Spatial Inform Sci XLIII-B3-2021:23–29. https://doi.org/10.5194/isprs-archives-XLIII-B3-2021-23-2021
7. Boak EH, Turner IL (2005) Shoreline definition and detection: a review. J Coast Res 21:688–703. https://doi.org/10.2112/03-0071.1
8. Bonnot-Courtois C, Levasseur JE, Denantes S (2013) Reconnaissance de la limite terrestre du domaine maritime : intérêt et potentialités de critères morpho-sédimentaires et botaniques. CETMEF
9. Campos-Taberner M, García-Haro FJ, Martínez B et al (2020) Understanding deep learning in land use classification based on Sentinel-2 time series. Sci Rep 10:17188. https://doi.org/10.1038/s41598-020-74215-5
10. Gibril MBA, Idrees MO, Shafri HZM, Yao K (2018) Integrative image segmentation optimization and machine learning approach for high quality land-use and land-cover mapping using multisource remote sensing data. JARS 12:016036. https://doi.org/10.1117/1.JRS.12.016036
11. Hegde AV, Akshaya BJ (2015) Shoreline transformation study of Karnataka coast: Geospatial approach. Aquatic Procedia, Int Conf Water Res Coastal Ocean Eng (ICWRCOE'15) 4:151–156. https://doi.org/10.1016/j.aqpro.2015.02.021
12. Hunt J (1959) Design of Seawalls and Breakwaters. J Waterways Harbors Division 85:123–152
13. LeCun Y, Bengio Y, Hinton G (2015) Deep learning. Nature 521:436–444. https://doi.org/10.1038/nature14539
14. Lecun Y, Bottou L, Bengio Y, Haffner P (1998) Gradient-based learning applied to document recognition. Proc IEEE 86:2278–2324. https://doi.org/10.1109/5.726791
15. Luijendijk A, Hagenaars G, Ranasinghe R, Baart F, Donchyts G, Aarninkhof S (2018) The state of the World's beaches. Sci Rep 8:6641. https://doi.org/10.1038/s41598-018-24630-6
16. Mallet C, Michot A, La Torre Y et al (2012) Synthèse de référence des techniques de suivi du trait de côte – Rapport. BRGM
17. Montavon G, Samek W, Müller K-R (2018) Methods for interpreting and understanding deep neural networks. Digit Signal Proc 73:1–15. https://doi.org/10.1016/j.dsp.2017.10.011
18. Niculescu S, Billey A Jr, H.T.-O.-A. (2018) Random forest classification using Sentinel-1 and Sentinel-2 series for vegetation monitoring in the Pays de Brest (France). In: Remote Sensing for Agriculture, Ecosystems, and Hydrology XX. Presented at the Remote Sensing for Agriculture, Ecosystems, and Hydrology XX, International Society for Optics and Photonics, p. 1078305. https://doi.org/10.1117/12.2325546
19. Robin M (2002) Télédétection et modélisation du trait de côte et de sa cinématique. In: Baron-Yelles N, Goeldner-Gionella L, Velut S (eds) Le littoral, regards, pratiques et savoirs. Etudes offertes à Fernand VERGER. Edition Rue d'Ulm/Presses universitaires de l'Ecole Normale Supérieure, Paris, pp 95–115
20. Toure S, Diop O, Kpalma K, Maiga AS (2019) Shoreline detection using optical remote sensing: a review. ISPRS Int J Geo Inf 8:75. https://doi.org/10.3390/ijgi8020075

Very High-Resolution Monitoring and Evaluation of Tidal and Ecological Restoration in Beaussais' Bay

Dorothée James, Antoine Collin, Antoine Mury, Mathilde Letard, O. Legal, and Alysson Lequilleuc

Keywords Tidal restoration · UAV · Diachronic monitoring · Multispectral · NDVI · NDWI · SAVI

1 Introduction

Since the Middle Ages, man has tried to expand the surface of his territory on the sea by draining the coastal marshes through the polderization process [1]. These recently-conquered lands were intended for agriculture and grazing. However, the current climate change and the latest projections of the IPCC (AR6) suggest an issue in this territorialization [4].

Tidal restoration is one of the possible mankind's adaptations facing the sea-level rise by letting the water recover its place [3].

The *Conservatoire du Littoral* is coordinating the European Life Adapto project (https://www.lifeadapto.eu/). The objective is to experiment flexible shoreline management on 10 experimental sites (nine in metropolitan France and one in French Guyana). Satellite or manned aerial monitoring of coastal ecosystems are widely

D. James (✉) · M. Letard · O. Legal · A. Lequilleuc
Ecole Pratique des Hautes Etudes – PSL University, Coastal GeoEcological Lab, Dinard, France
e-mail: dorothee.james@ephe.psl.eu; mathilde.letard@ephe.psl.eu; alysson.lequilleuc@ephe.psl.eu

A. Collin
Ecole Pratique des Hautes Etudes – PSL University, Coastal GeoEcological Lab, Dinard, France

LabEx CORAIL, Papetoai, French Polynesia
e-mail: antoine.collin@ephe.psl.eu

A. Mury
UMR AMAP, IRD, CIRAD, CNRS, INRAE, Université de Montpellier, Montpellier, France
e-mail: antoine.mury@ird.fr

S. Niculescu (ed.), *European Spatial Data for Coastal and Marine Remote Sensing*, https://doi.org/10.1007/978-3-031-16213-8_12

used for their continuous spatial coverage (several hectares or square kilometers) and their large spectrum [2].

The monitoring of the coastline by unmanned aerial vehicles (UAV) is in a growing trend due to its ease of deployment [6]. The UAV allows to monitor a coastal ecosystem with a very high temporality. However, red-green-blue (RGB) cameras equipping UAV limit the possibilities of discrimination of the landscape coverage like soil or vegetation [5].

This original study proposes to evaluate the evolution of a polder within the framework of the flexible management of the coastline by UAV. Using multispectral (RGB, red edge [RE], and near infrared [NIR]) spectral wavebands at very high resolution, we examine how significant the variation of the landscape coverage will be detected using geospatial indices.

2 Methodology

Beaussais' polder is located at the bottom of Beaussais' bay on the French Emerald Coast (48°35'N; 2°09'W). This land reclaimed from the sea is protected from the impact of the marine weather by a vegetated dike in front of which a salt meadow has developed. The central part of the polder is anthropized and exploited as agricultural and grazing land. The polderization of the Beaussais' site began in the sixteenth century with the construction of the *Digues aux Moines* and the draining of the marsh. The assaults of the waves have recently (March 2020) created a breach in the dike allowing the sea to enter the polder.

Three UAV planned surveys and ground-truth campaigns have been conducted on February 19, 2021, on September 3, 2021, and on February 8, 2022. A DJI Phantom 4 (P4PROV2) equipped with a 20 M-pixel RGB camera (4864 × 3648 pixels) acquired 1415 photographs. A 1.2 M-pixel Parrot Sequoia+ multispectral sensor (1280 × 960 pixels for each camera; G, R, RE, NIR) completed the acquisition with 8165 photographs collected.

Simultaneously to the aerial acquisitions, nine ground control points (GCPs) were positioned on the polder and collected with a Topcon centimetric D-GNSS Hiper V and a Topcon datacontroller FC 500.

From the aerial images, RGB, RE and NIR orthomosaics were generated by photogrammetry using the Pix4D software at 0.1 m and 0.06 m pixel size, respectively.

Orthomosaics were individually georeferenced from the nine GCPs in ArcGIS software and were stacked on one single layer.

The stacked RGB, RE and NIR orthomosaics consisted of the foundation for the calculation of land cover index (Fig. 1): normalized difference vegetation index (NDVI), normalized difference water index (NDWI), and soil-adjusted vegetation index (SAVI).

At the study site scale, indices' variances were computed between the three periods to map the evolution.

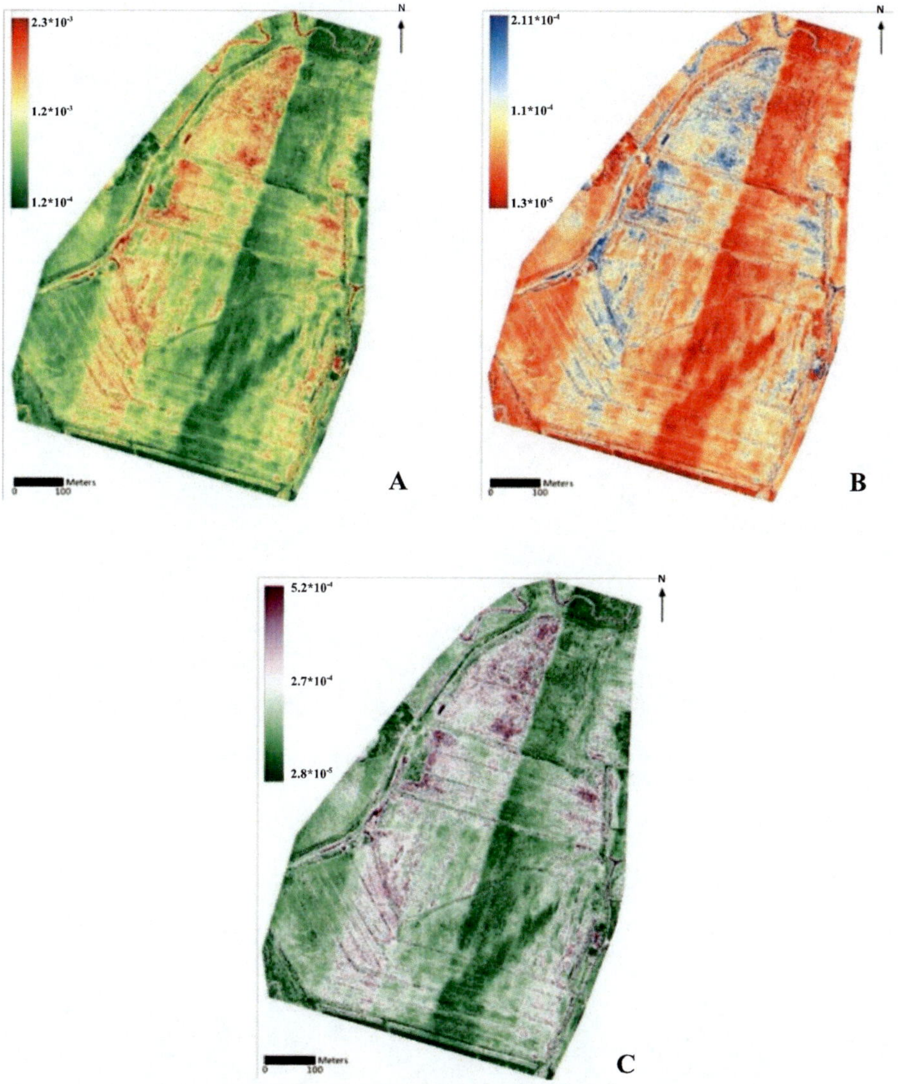

Fig. 1 (**a**) Normalized difference vegetation index (NDVI) variance, (**b**) normalized difference water index (NDWI) variance, and (**c**) soil-adjusted vegetation index (SAVI) variance

At the local scale, a transect was extracted to finely assess the evolution of 11 stations along a transect line for each land cover index and for each campaign according to a west–east gradient, from sea to land (Fig. 2). Standard deviations and means were computed and compared for NDVI (Fig. 3).

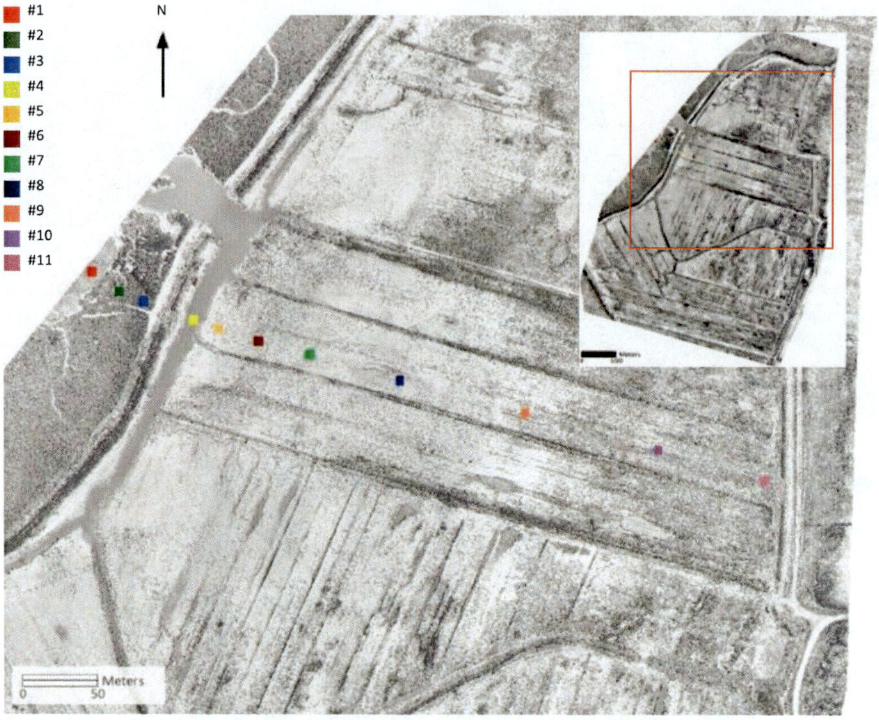

Fig. 2 Zoom of the location of the 11 stations extracted along a west–east transect

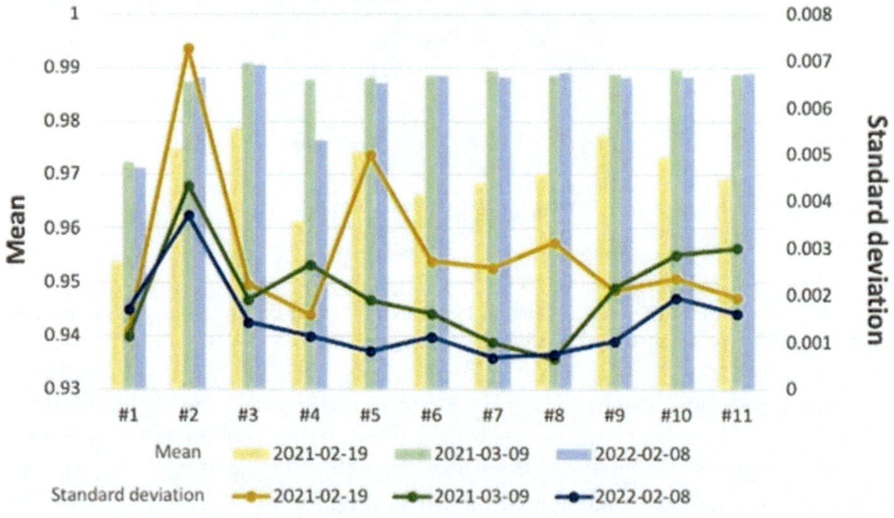

Fig. 3 NDVI means and standard deviations for the three campaigns

3 Results and Discussion

The variance of the three land cover indices (NDVI, NDWI and SAVI) indicates changes in vegetation and mineral cover of the polder (Fig. 1). Given the variance measures the statistical distance of the indices from the mean, a high variance value indicates a high dispersion of the pixels from the mean for the three campaigns.

The polder land cover has been modified by the successive marine submersion through the increasingly large breach. Furthermore, at the scale of the NDVI classes, the standard deviation is higher within the same class for the February 2021 campaign. For the other two dates, the pixel values are less scattered. Two hypotheses can explain this phenomenon: the phenology and the modification of the vegetation types (halophytic vegetation vs. non-halophytic vegetation).

Acknowledgments The authors thank the *Conservatoire du littoral* for their funding of the Life Adapto project and their confidence in the project.

References

1. Collin A, Durozier T, Hervouet G, Costa S (2021) Détecter le processus de dépoldérisation par l'observation satellite à très haute résolution temporelle. SAGEO 2021, May 2021, La Rochelle, France
2. Collin A, Etienne S, Feunteun E (2017) VHR coastal bathymetry using WorldView-3 : colour versus learner. Remote Sensing Letters 8(11):1072–1081
3. Lydie G-G, Verger F (2009) Du "polder" à la "dépoldérisation"? L'espace géographique 38(4):376–377
4. IIPCC (2022) Climate change 2022: impacts, adaptation, and vulnerability. Contribution of working group II to the sixth assessment report of the intergovernmental panel on climate change. Cambridge University Press. In Press
5. James D, Collin A, Mury A, Letard M, Guillot B (2021) UAV multispectral optical contribution to coastal 3D modelling. IEEE International Geoscience and Remote Sensing Symposium IGARSS. pp 951–7954
6. Mury A, Collin A, James D (2019) Morpho–sedimentary monitoring in a coastal area, from 1D to 2.5 D, using airborne drone imagery. Drones 3(3):62

Assessment of Shoreline Change of Jerba Island Based on Remote Sensing Data and GIS Using DSAS Tools

Amina Boussetta, Simona Niculescu, Soumia Bengoufa, and Mohamed Faouzi Zagrarni

Keywords Remote sensing · OBIA · PBIA · Band ratioing · CNN · Jerba

1 Introduction

Sandy coasts are often marked by an intense recession of the shoreline [1]. Recent advances in the radiometric, spatial, temporal, and spectral resolution of sensors have provided a valuable tool set for innovative coastal data processing methods. It has been demonstrated that satellite imagery, as well as new remote sensing methods, can provide more practical approaches to the mapping and monitoring of coastal environments.

The evolution, or rate of change, of the coastline is an important parameter that indicates the accretion/erosion pattern in a coastal area [3]. The calculation of historical dynamics and monitoring of shoreline change in the studied area were performed using data captured by remote sensing methods. In this regard, this study highlights the different image processing techniques, including image classification, which can be applied to study the coastal environment. Machine learning and deep learning algorithms, such as the neural networks and random forest algorithm,

A. Boussetta (✉) · M. F. Zagrarni
U.R Applied Hydrosciences UR13ES81, Higher Institute of Water Sciences and Techniques, University of Gabes, Cité Erryadh, Gabes, Tunisia

S. Niculescu
Laboratory LETG-Brest, UMR 6554 CNRS, University of Western Brittany, Plouzané, France
e-mail: simona.niculescu@univ-brest.fr

S. Bengoufa
University of Western Brittany, CNRS, LETG Brest UMR 6554 CNRS, Technopôle Brest-Iroise, Av. Dumont d' Urville, Plouzané-Brest, France

National School of Marine Sciences and Coastal Management (ENSSMAL), University Campus of Dely Ibrahim Bois des Cars, Algiers, Algeria

S. Niculescu (ed.), *European Spatial Data for Coastal and Marine Remote Sensing*, https://doi.org/10.1007/978-3-031-16213-8_13

have recently been successfully adopted for remote sensing applications [5, 6]. However, Toure et al. [7] claimed that the suitability of these methods for shoreline detection has not yet been sufficiently studied. This work focuses on the detection, extraction, and historical study of coastline positions from satellite data processing. The use of Geographic Information System (GIS) tools, and the geostatistical calculation tool Digital Shoreline Analysis System (DSAS), enables the quantification of erosion and accretion rates and facilitates the description of evolutionary trends in our study area over 32 years, covering three periods. This study aims to validate precise techniques for monitoring sandy coasts.

2 Methodology

In the Gulf of Gabes, Jerba, the largest island in North Africa, is an island of 514 km^2 with 150 km of coastline. It is located in the southeast of Tunisia at a northern latitude between 33°57′0.56" and 33°37’46.82" and an eastern longitude between 10°45′38.43" and 11°3’53.802" (Fig. 1). The island of Jerba boasts the most beautiful sandy beaches, which gives it significant recreational interest. The eastern coast

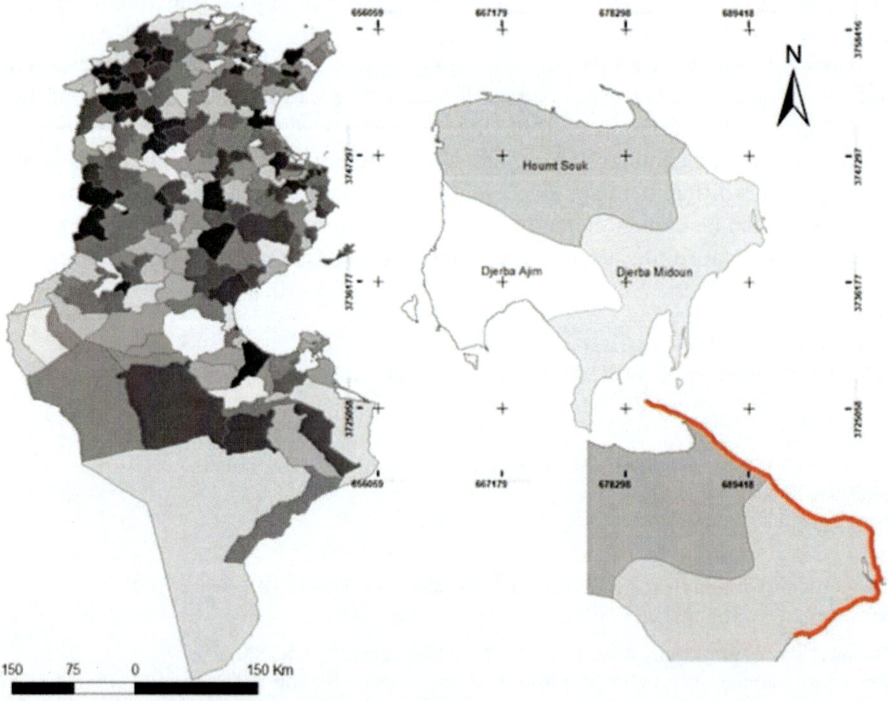

Fig. 1 Geographic location of the study area

of the island of Jerba has long been the subject of intensive touristic development due to its attractive natural potential.

In order to extract shoreline features from satellite images (Fig. 1), the band ratioing method, an integrated *convolutional neural network* (CNN) method, and the development of machine learning algorithms were the techniques used in this study. Using the digital shoreline assessment system (DSAS), statistical parameters were derived to estimate erosion rates. Two types of satellite data were used for the extraction and diachronic analysis of the shoreline evolution along the sandy beaches of Jerba Island.

Ten Landsat images were used, downloaded free of charge in GeoTIFF format via the USGS Earth Explorer website (http://www.earthexplorer.usgs.org). As well, two Sentinel-2 images with a processing level of 2A and 2B were downloaded via their platform (Muscat Distribution Workshop). For this work, the images with the "flat reflectance" band were chosen. All the satellite images have zero cloud cover.

Supervised classification requires a user to instruct the system by designating areas of the image as representative samples of the classes to be extracted. The classification is thus preceded by a learning process. In order to instruct the system, it is necessary to have a good knowledge of the land use. For this, we need ground-truth data. A ground truth is a set of guaranteed exact data on the observed land use. It is achieved by close observations in the field. The classes are as follows: for the Sentinel image: (1) water, (2) wet sand, (3) dry sand, (4) urban, and (5) vegetation; for the Landsat image: (1) water, (2) sand, (3) vegetation, and (4) others.

In a GIS, the shoreline indicator is digitized and then overlaid to compare its position with other older shorelines. Rates of change can be calculated automatically with special tools such as the DSAS. Three methods in particular were applied: *end point rates* (EPR), *linear regression* (LR), and *net shoreline movement* (NSM), which represents the metric distances between two shorelines (Fig. 2) [4].

3 Results and Discussion

The knowledge and monitoring of the shoreline and/or the evolution of the coastal fringe is an essential element in the prevention of coastal hazards [2] and the assessment of the vulnerability of our study area. As such, it is critical to use data with high resolution and employ good accuracy techniques in order to obtain relevant results. The (semi-) automatic extraction results were then compared to those obtained by manual digitizing of the coastline considered as reference line (Fig. 3). The OBIA-RF algorithm better corresponds to the in situ measurements and appears to have fewer ambiguities in the classifications and fewer errors for the wet soil class: a fundamental class considered as an essential indicator for the detection of the shorelines of sandy beaches (Fig. 2). The results show a general trend of erosion (89% for the coastline). The statistics revealed that the study area, with the exception of the sandy spit, has experienced an intense and alarming level of erosion.

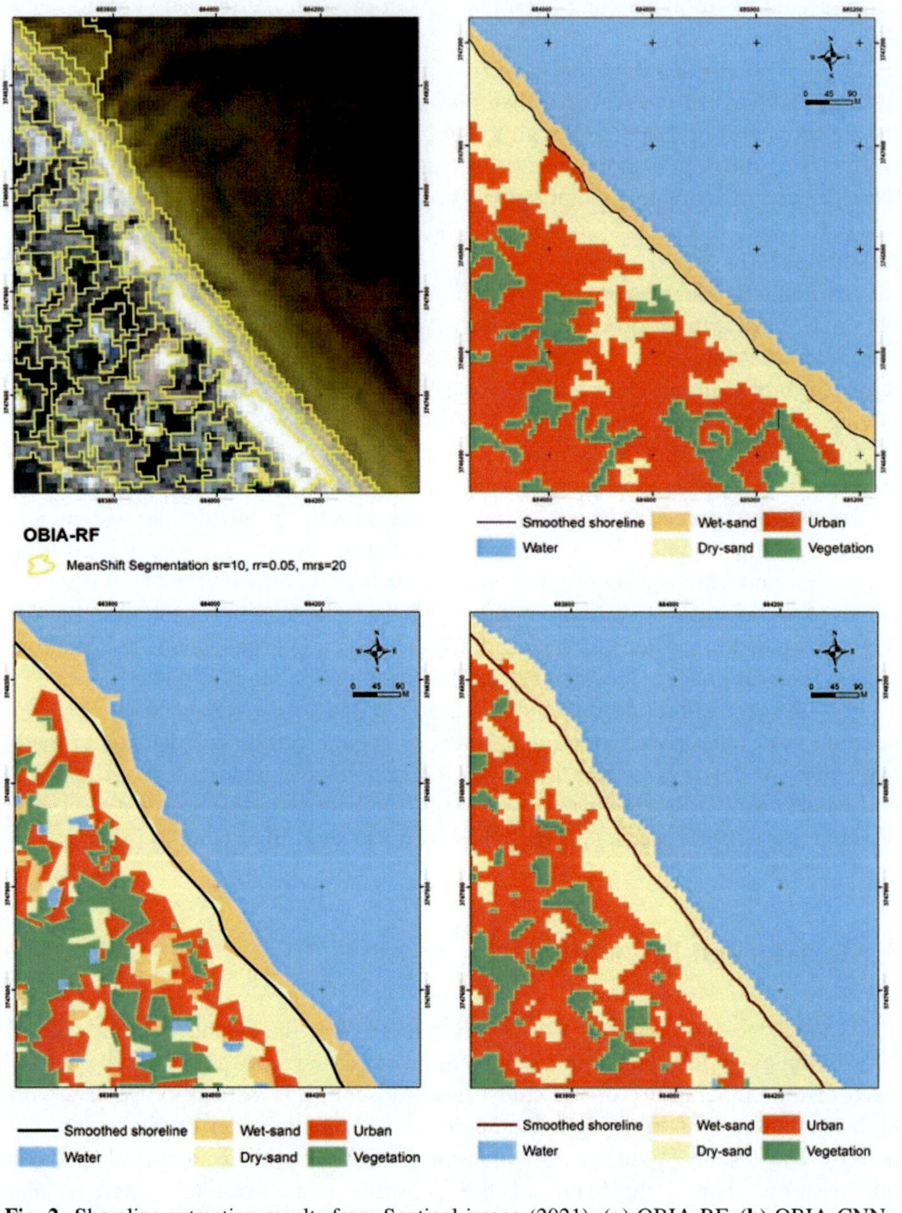

Fig. 2 Shoreline extraction results from Sentinel image (2021): (**a**) OBIA-RF, (**b**) OBIA-CNN, (**c**) PBIA-RF

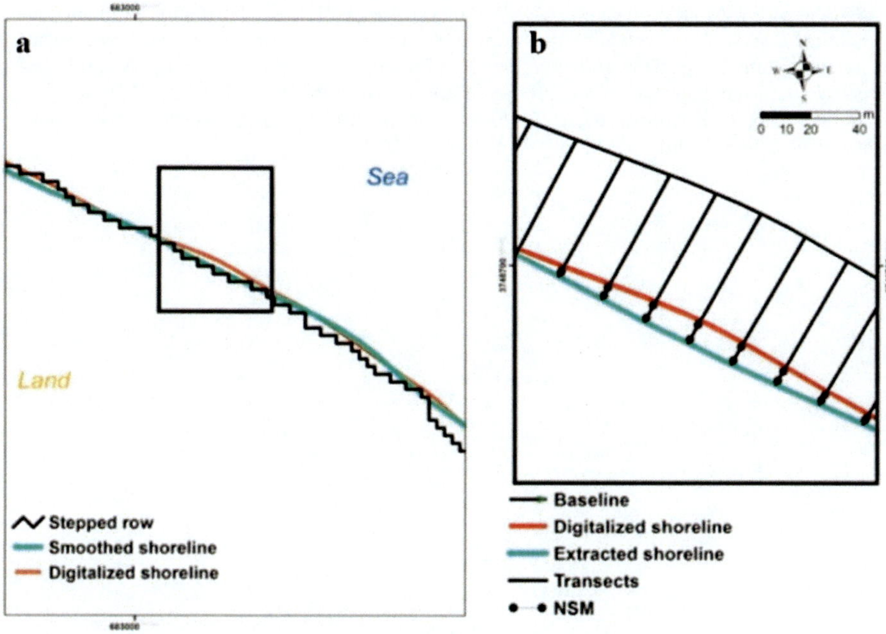

Fig. 3 (**a**) Shoreline smoothing and (**b**) transect generation and NSM

Acknowledgments The authors thank the ERASMUS+ program, the Jean Monnet Chair, and the European Spatial Studies of Sea and Coastal Zones -599967-EPP-1-2018-1-FR-EPPJMO-CHAIR.

References

1. Amrouni O, Hzami A, Heggy E (2019) Photogrammetric assessment of shoreline retreat in North Africa: anthropogenic and natural drivers. ISPRS J Photogramm Remote Sens 157:73–92. https://doi.org/10.1016/j.isprsjprs.2019.09.001
2. Bengoufa S, Niculescu S, Mihoubi MK, Belkessa R, Rami A, Rabehi W, Abbad K (2021) Machine learning and shoreline monitoring using optical satellite images: case study of the Mostaganem shoreline. Algeria J Appl Rem Sens 15. https://doi.org/10.1117/1.JRS.15.026509
3. Boumboulis V, Apostolopoulos D, Depountis N, Nikolakopoulos K (2021) The importance of geotechnical evaluation and shoreline evolution in coastal vulnerability index calculations. JMSE 9:423. https://doi.org/10.3390/jmse9040423
4. Himmelstoss EA, Henderson RE, Kratzmann MG, Farris AS (2018) Digital shoreline analysis system (DSAS) version 5.0 user guide (USGS numbered series no. 2018–1179), digital shoreline analysis system (DSAS) version 5.0 user guide, open-file report. U.S. Geological Survey, Reston, VA https://doi.org/10.3133/ofr20181179
5. Niculescu S, Xia J, Roberts D, Billey A (2020) Rotation forests and random forest classifiers for monitoring of vegetation in pays De Brest (France). Int Arch Photogramm Remote Sens Spatial Inf Sci XLIII-B3-2020:727–732. https://doi.org/10.5194/isprs-archives-XLIII-B3-2020-727-2020

6. Sr SN, Billey A, Jr, H.T.-O.-A. (2018) Random forest classification using Sentinel-1 and Sentinel-2 series for vegetation monitoring in the Pays de Brest (France). In: Remote sensing for agriculture, ecosystems, and hydrology XX. Presented at the remote sensing for agriculture, ecosystems, and hydrology XX, SPIE, p. 1078305. https://doi.org/10.1117/12.2325546
7. Toure S, Diop O, Kpalma K, Maiga A (2019) Shoreline detection using optical remote sensing: a review. IJGI 8:75. https://doi.org/10.3390/ijgi8020075

New Insights into the Shallow Morpho-Sedimentary Patterns Using High-Resolution Topo-Bathymetric Lidar: The Case Study of the Bay of Saint-Malo

Bruno Caline, Antoine Collin, Yves Pastol, Mathilde Letard, and Eric Feunteun

Abstract A detailed morpho-sedimentological map of this bay, derived from the calibrated lidar rasterization with field data, provides new insights into the relationships between the nature and morphology of the sediment bodies and their overall distribution within this megatidal bay protected by numerous islands and rocky shoals.

Keywords Sedimentology · Geomorphology · Macrotidal environment · High-resolution mapping · Marine geomorphometry

1 Introduction

Ocean–climate changes entail rapid movements of sedimentary stocks along the coast that need to be well understood and mapped at high resolution to cope with the shoreline erosion and marine submersion [3].

B. Caline (✉)
Sedimentologist, Dinard, France

A. Collin · M. Letard
Ecole Pratique des Hautes Etudes – PSL Université, LETG-Dinard UMR 6554, Dinard, France
e-mail: antoine.collin@ephe.psl.eu; mathilde.letard@ephe.psl.eu

Y. Pastol
Service Hydrographique et Océanographique de la Marine, Brest, France
e-mail: yves.pastol@shom.fr

E. Feunteun
Museum National d'Histoire Naturelle, CNRS BOREA, Dinard, France
e-mail: eric.feunteun@mnhn.fr

S. Niculescu (ed.), *European Spatial Data for Coastal and Marine Remote Sensing*, https://doi.org/10.1007/978-3-031-16213-8_14

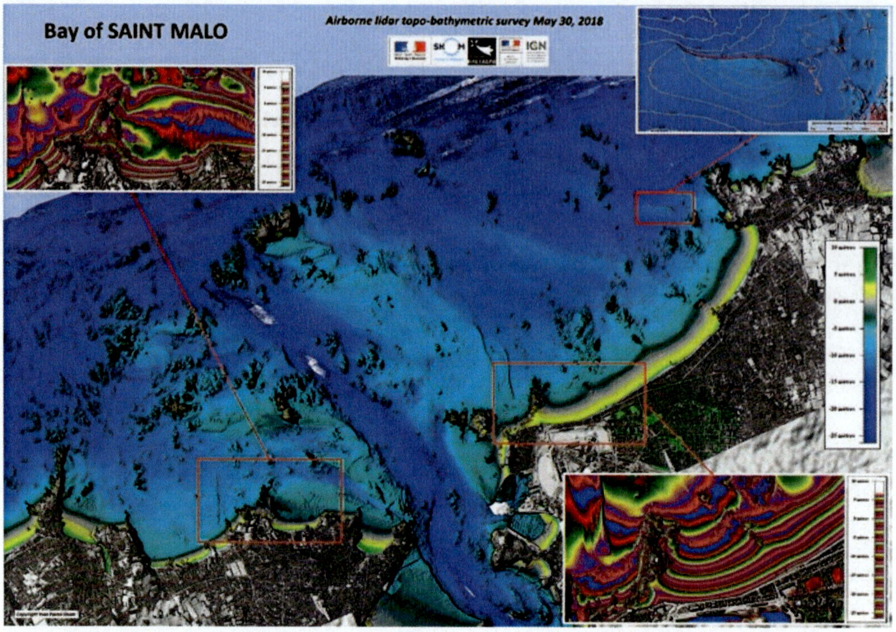

Fig. 1 Airborne lidar topo-bathymetric survey of the Bay of Saint-Malo (May 30, 2018)

The mapping of the coastal seabed is conventionally ensured by acoustic surveys, provided with multibeam and side-scan sonars. This kind of survey can be strongly impeded by shallow depths, which make the navigation dangerous even impossible [6]. Recently, satellite-derived bathymetry has proven to be useful and accurate enough to retrieve both seabed topography and composition [4]. However, this regional technique but fine-scale technique is greatly constrained by the water clarity.

Overcoming the active sound and the passive light surveys, the airborne lidar system provided with a topo-bathymetric sensor has the capabilities to seamlessly scan the land–sea continuum at very high resolution at three times the Secchi depth [5].

We propose a novel coastal characterization of the complex but representative morpho-sedimentary seabed patterns using the latest very-high-resolution topo-bathymetric lidar data, its morphometric derivatives, and sonar and sediment samples across the Bay of Saint-Malo (Fig. 1).

2 Study Site

The study site is located in the southern sector of the Normandy–Brittany Gulf in the English Channel. As part of the *Côte d'Emeraude* shoreline in Brittany, the studied area is centered in the walled city of Saint-Malo and extends from the *Pointe du Décollé* in the west to the *Pointe de la Varde* in the East and includes the *Cézembre* Island in the north. This bay is characterized by a number of physical and biological specificities, which result in a unique marine environment along the Brittany coast: a megatidal regime with tidal range reaching 13.5 m, very shallow waters (less than 30 m deep) with numerous islands and rocky headlands, strong tidal currents and major wave action controlled by dominantly western winds, numerous sediment types due to a wide range in grainsize and carbonate content [1], the occurrence of an incised valley inherited to the paleo-Rance river, various benthic habitats that result in high biodiversity, and well-developed sand shoals and bars with various bedforms generated by high hydrodynamics.

3 Topo-Bathymetric Lidar Acquisition

The topo-bathymetry acquisition, piloted by the Shom (French hydrographic office), took place during the Brittany survey of the bays of Mont-Saint-Michel and Saint-Malo. The campaign from *Pointe du Grouin* to Saint-Briac was carried out on May 30, 2018. The system used for the survey was a HawkEye III (Sweden) combining three lidars ("topo," "shallow," and "deep" sensors) and two cameras. The "topo" is used for the topographic area and the sea surface. The "shallow" works less than 10 m deep and the "deep" sensor can achieve 50 m in very clear waters. During this outing, 25 m deep was reached. The full coverage of the area was very satisfactory (only two gaps in the *La Petite Porte* channel). Topographic data had a density of 12 pts/m^2. Bathymetric shallow and deep data reached 1.5 pt/m^2 and 0.4 pt/m^2, respectively. The data acquisition and processing were carried out by the Shom coastal altimetry team as part of the Litto3D® project [7], with the RGF93 datum, projected on Lambert93, and vertically referenced to the IGN69.

4 Sonar and Sediment Samples

In addition to the newly acquired topo-bathymetric lidar data of the bay of Saint-Malo, field acquisition including echosounder and side-scan sonar profiles, seabed photography and sediment dredging have been carried out on May 2021.

Calibrating the lidar prediction, those ancillary datasets allow revisiting the morphology and nature of the seabed within this highly variable depositional environment. The main depositional bodies investigated are: shallow water sand shoals, deeper water pebbly channels, wide rocky substrate, and sheet-like accumulation of red algae (maërl beds).

5 Preliminary Results

The bay of Saint-Malo is featured with a unique depositional model for cold temperate shallow waters [2]. This new facies model highlights the high percentage of carbonate grains in the sediments due to the enhanced biodiversity of the benthic habitats. A detailed morpho-sedimentological map of this bay, derived from the calibrated lidar rasterization with field data, provides new insights into the relationships between the nature and morphology of the sediment bodies and their overall distribution within this megatidal bay protected by numerous islands and rocky shoals.

Acknowledgments The authors gratefully acknowledge Loïc Le Goff, Angéline Bulot, Antoine Mury, and Julien Guillaudeau for their waterborne support.

References

1. Augris C, Blanchard M, Bonnot-Courtois C, Houlgatte E, Le Vot M (2000) Carte des formations superficielles sous-marines entre le Cap Fréhel et Saint-Malo – échelle 1/20 000. Ifremer
2. Caline B, Pariente M, Desroy N, Le Mao P, Rollet C (2018) A spot of enhanced biodiversity and carbonate production in the cold temperate shallow waters of the English Channel: the Normandy-Brittany Gulf. In International Sedimentological Congress, 13–17 August 2018, Québec, Canada
3. Collin A, Duvat V, Pillet V, Salvat B, James D (2018) Understanding interactions between shoreline changes and reef outer slope morphometry on Takapoto Atoll (French Polynesia). J Coast Res 85:496–500
4. Collin A, Laporte J, Koetz B, Martin-Lauzer FR, Desnos YL (2020) Coral reefs in Fatu Huku Island, Marquesas Archipelago, French Polynesia. In: Seafloor geomorphology as benthic habitat. pp 533–543. Elsevier
5. Collin A, Long B, Archambault P (2012) Merging land-marine realms: spatial patterns of seamless coastal habitats using a multispectral LiDAR. Remote Sens Environ 123:390–399
6. Kenny AJ, Sotheran I (2013) Characterising the physical properties of seabed habitats. Methods for the study of marine benthos, pp 47–95
7. Shom Litto3D® https://www.diffusion.shom.fr

Spatial Modeling of the Benthic Biodiversity Using Topo-Bathymetric Lidar and Neural Networks

Angéline Bulot, Antoine Collin, Mathilde Letard, Eric Feunteun, Loic Le Goff, Yves Pastol, and Bruno Caline

Keywords Coastal mapping · Shannon diversity index · Non-linear regression · Bay of Saint-Malo

1 Introduction

The rich biodiversity of bays and estuary areas provides numerous services to human populations including food support, agricultural amendments [3], and tourism. These ecosystems are also the first players in coastal protection and erosion control [1]. In a global climate change context, the loss of biodiversity is critically depleting these ecosystem contributions. On the Channel French coast, bays and estuaries show large sediment cover variations especially in unique human-modified areas like the Rance estuary [5]. Calculating a discrete Shannon index eases the modelling of the benthic diversity by quantifying the proportion of each biological and geological class. Accurate descriptions of coastal basin structures are thus proposed using Shannon index evaluations [4].

A. Bulot (✉) · A. Collin · M. Letard
Ecole Pratique des Hautes Etudes – PSL Université, LETG-Dinard UMR 6554, Dinard, France
e-mail: antoine.collin@ephe.psl.eu; mathilde.letard@ephe.psl.eu

E. Feunteun · L. Le Goff
Museum National d'Histoire Naturelle, CNRS BOREA, Dinard, France
e-mail: eric.feunteun@mnhn.fr; loic.le-goff1@mnhn.fr

Y. Pastol
Service Hydrographique et Océanographique de la Marine, Brest, France
e-mail: yves.pastol@shom.fr

B. Caline
Sedimentologist Expert, Dinard, France

S. Niculescu (ed.), *European Spatial Data for Coastal and Marine Remote Sensing*, https://doi.org/10.1007/978-3-031-16213-8_15

Sonar and satellite imagery techniques enable the mapping of coastal areas, which are accessible by boat and where turbidity remains low, respectively [2]. The very-high-resolution (VHR) bathymetric lidar technique offers a more precise description of the shallow and deep seabed coverage, shedding light on regional benthic areas.

The present study aims at innovatively mapping a whole benthic geobiodiversity Shannon index, by means of morphometric continuous predictors derived from a combination of VHR topo-bathymetric lidar data, underwater imagery discrete response, and a 10-neuroned neural network prediction.

2 Methodology

The study is taking place in the Bay of Saint-Malo, along the coast from Saint-Briac-sur-Mer (48°38.196 N, 2°08.132 W) to *La Pointe du Grouin* (48°41.846 N, 1°53.940 W). Six marine imagery missions enabled access to a total of 212 location points in May 2021 (Fig. 1). At each sampling point, the GPS location was indicated by a Garmin eTrex 10 and an underwater video was acquired with a HERO 3+ GOPRO camera, fixed on a 25 × 25 cm^2 photoquadrat. The camera was lowered until the photoquadrat landed on the seabed. The topo-bathymetric airborne lidar campaigns (May, 2, 2018 – July, 4, 2018) enabled the bathymetric measurement and the seafloor habitat mapping of the bay. The Leica HawkEye III lidar tool provided a digital terrain model (DTM) derived from a point cloud with a spatial resolution of 2 m. The density of the point cloud was 0.4 point/m^2 for the deep channel and 1 points/m^2 for the shallow channel. Establishing a geobiological classification (Table 1) of all the structures observed inside the photoquadrat allowed calculation of the global Shannon index of the marine area (Eq. 1).

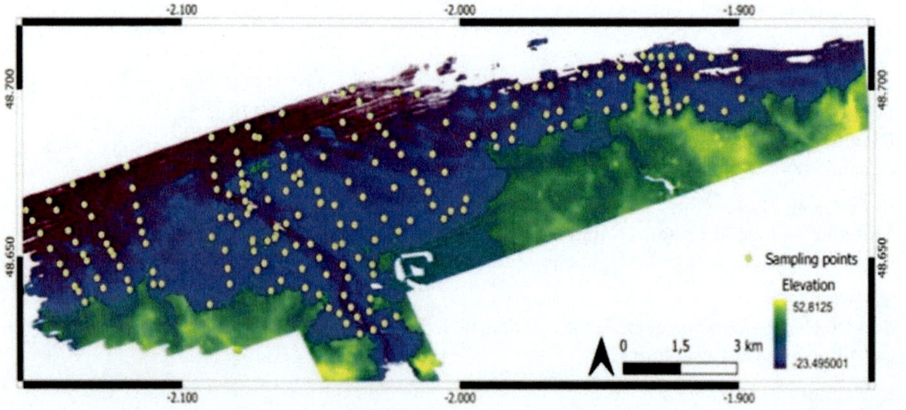

Fig. 1 Study area: the Bay of Saint-Malo, France

Table 1 Biological and geological classes used in the study

Geological class name	Photoquadrat	Biological class name	Photoquadrat
Boulder		Seaweed	
Cobble–pebble		Seagrass	
Gravel		Shells	
Sand	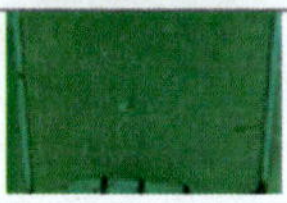		

$$I = -\sum_{i} (x_i / X)^* \ln (x_i / X) \quad (1)$$

where I is the Shannon geobiodiversity index, x_i is the proportion of each class, and X is the total proportion of the geobiological classes.

The biological diversity and the geological diversity were also computed by calculating the proportions of biological and geological classes. With a dataset of 19 morphometric features, linear algorithms and non-linear algorithms, including neural networks, were tested to find the model that best describes the distribution of the global geobiodiversity index within the bay of Saint-Malo. The number of lidar-derived features used, the number of hidden neurons (with hyperbolic tangent activation function), and the number of tests the algorithm performed were the parameters used to refine the models. The prediction models were either discarded or retained depending on their coefficient of determination R^2.

3 Results and Discussion

With the maximum of 19 lidar parameters involved, linear regression models fail to predict the geobiodiversity with an averaged R^2 of 0.15, whereas non-linear partition models offer a much better prediction with an averaged R^2 of 0.70. In the case of neural network algorithms, increasing the number of tests moderately enhances the previous propositions with an average R^2 of 0.76. Increasing the number of hidden neurons gives the more satisfactory model: while the prediction with a single neuron has an averaged R^2 of 0.23, it becomes really accurate with 10 neurons (R^2 = 0.98). The same models with the sole biodiversity or geodiversity Shannon

Fig. 2 Map of benthic geobiodiversity of the bay of Saint-Malo

index suggest the prediction models are better with the geodiversity index values. Some of the lidar features have a bigger influence on the Shannon index predictions such as elevation, ruggedness, valley depth, and exposition: with only these four features, neural network models are able to calculate the geobiodiversity index with an average R^2 coefficient of 0.55. However, maximizing the number of morphometric parameters clearly refines the predictions: for linear and non-linear models, the accuracy of the predicted values distinctly increases with the number of predictors. Using the best hyperbolic tangent 10-neuron model with all the 19 morphometric parameters, an original map of the current benthic geobiodiversity of the bay of Saint-Malo was built (Fig. 2). It offers an overview of the spatial distribution of the geological and biological structures.

Further studies based on Shannon index should be based on a wider range of geobiological classes. In this work, the limited number of biological structures may explain the less satisfactory predictions of the biodiversity index. Better results could be obtained by considering species of seaweed individually as opposed to being grouped together. The same goes for seagrasses. Finally, it would be useful to sample the seabed at several depths: the geobiodiversity sample inside the photo-quadrat was sometimes not representative enough of the surrounding environment. Capturing and classifying the habitats at 15 m height from the seafloor is likely to better match with the underwater imagery geolocation (Table 2).

Table 2 Name and description of the morphometric parameters used in the study

Name	Definition	Name	Definition
Angle	Angle between the topographic surface and the plane defined by cutting lines	Relief	Values of the 3 pixel colors calculated with landform values
CLdip	Angle of cutting lines dip	Ruggedness	Index of terrain heterogeneity
CLdipdir	Angle of cutting line dip direction	Slope	Angle to the horizontal, on the whole segment
Elevation	Value in the numerical terrain model	Slope height	Distance from top to bottom of a steep landform
Exposition	Angle between the surface and incoming light	Slope position	Relative position of steep landforms
Half slope	Angle to the horizontal, on the half segment	Shading	
Landforms	Index of topographic type areas	TPI	Value of gray pixel scale
MRRTF	Index of ridge top flatness	Valley depth	Index of topographic position
MRVBF	Index of valley bottom flatness		Distance from the bottom of a valley to the top of a base channel network
Normalized height	Altitude value bounded by the altitude above drain culmination and below summit culmination	Wind exposition	Index of windward situation

References

1. Barbier EB, Hacker SD, Kennedy C, Koch EW, Stier AC, Silliman CR (2011) The value of estuarine and coastal ecosystem services. Ecol Monogr 81:169–193
2. Collin A, Laporte J, Koetz B, Martin-Lauzer FR, Desnos YL (2016) Mapping bathymetry, habitat, and potential bleaching of coral reefs using Sentinel-2. In: 13th International Coral Reef Symposium (ICRS 2016), pp 405–420
3. Ferec C, Chauvin T (1985) The exploitation of aggregates and marine amendments in the Normandy-Brittany Gulf. IFREMER report 85/2430003 4:7–77
4. Foveau A (2009) Macrobenthic habitats and communities in the eastern english channel : a state of the art study and survey at the beginning on the XXith century. Lille Nord de France University
5. Jigorel A, Ledrappier M (2003) Sedimentary and biological impacts of the managements of sediments in the marine basin of the Rance. pp 163–177

Local Circalittoral Rocky Seascape Structuring Fish Community: Insights from a Photogrammetric Approach

Quentin Ternon, Antoine Collin, Eric Feunteun, Frédéric Ysnel, Valentin Danet, Manon Guillaume, and Pierre Thiriet

Keywords Fish community · Seascape · Photogrammetry · Architecture · Substrate composition · Benthic community

1 Introduction

Fish face multiple environmental pressures acting as multiscale filters structuring the community [6]. Study of the effect of local (~100 m^2) habitat components such as habitat architecture, substrate composition, and benthic community on fish community is still limited because of the technical difficulties to sample reliable

Q. Ternon (✉) · E. Feunteun
Museum National d'Histoire Naturelle (MNHN), Station Marine de Dinard, Centre de Recherche et d'Enseignement sur les Systèmes CÔtiers (CRESCO), Dinard, France

Laboratoire Biologie des Organismes et Écosystèmes Aquatiques (UMR BOREA), MNHN, CNRS, Sorbonne Université, Université de Caen Normandie, Université des Antilles, IRD, Paris Cedex, France
e-mail: quentin.ternon@mnhn.fr; eric.feunteun@mnhn.fr

A. Collin
Ecole Pratique des Hautes Etudes – PSL University, Coastal GeoEcological Lab, Dinard, France
e-mail: antoine.collin@ephe.psl.eu

F. Ysnel
Université de Rennes 1, UMR BOREA, MNHN, Rennes Cedex, France

V. Danet · M. Guillaume
Museum National d'Histoire Naturelle (MNHN), Station Marine de Dinard, Centre de Recherche et d'Enseignement sur les Systèmes CÔtiers (CRESCO), Dinard, France

P. Thiriet
Patrimoine Naturel (UAR OFB–CNRS–MNHN), MNHN – Station Marine de Dinard, Dinard, France

S. Niculescu (ed.), *European Spatial Data for Coastal and Marine Remote Sensing*, https://doi.org/10.1007/978-3-031-16213-8_16

descriptors of all these habitat components. Nevertheless, the effect of the 3D architecture, especially the complexity, has been highlighted to act as an important variable, locally structuring fish communities and leading to an increase in diversity (species richness and Shannon index) and quantity (total density and biomass) of fish [1]. Photogrammetry is extending to submarine environment and allows to produce very fine information of the architecture and substrate composition [5], as well as of the benthic community [4]. The influence of the different habitat components (i.e., architecture, substrate composition, and benthic community) on the structure of fish community is here investigated, as well as the interest of photogrammetry in comparison to visual observation (Fig. 1).

2 Methodology

The sampling area is located in the Saint-Malo's bay (Brittany, France). Tree distinct coastal rocky reefs (2.5–4.5 nautical miles apart from each other) have been sampled by scuba diving with 2 target depths: −8 and −13 corrected at the lowest astronomical tide level. Fish communities and habitats were described on 120 m^2 georeferenced transects (30 m long, 4 m wide) subdivided in 3 successive sections of 40 m^2. Two replicates (transects) were sampled when possible, for each depth on each site, totaling 12 transects (spatial sampling units).

Fish communities were described through visual census, noting the abundance, size, and species of each demersal and cryptobenthic fish. Four temporal replicates were carried out (2 in June–July 2020 and 2 in September–October 2020) returning to the same transect, thanks to visual signposts that were deployed.

Habitat was sampled only once (period). Habitat components were categorized into three distinct components, each of which being described by photogrammetry and visual observation (Fig. 1). Architecture (Fig. 1a) was described by various geomorphological descriptors computed from the digital surface model (DSM) produced by the 3D photogrammetric model or visually estimating the slope or degree of complexity (score from 1, flat, to 4, very complex) in 2 m × 2 m quadrats. The substrate composition (Fig. 1b) was mapped through a supervised classification (8 typological classes: bedrock, big/medium/small boulder, pebble, gravel, sand, artificial [5]) of the photogrammetric model allowing calculation of various landscape descriptors. Visual observation enabled listing the different substrate typologies (the same classes as by photogrammetry) in 2 m × 2 m quadrats. Benthic community was sampled with a process of stratified (by typological classes) virtual quadrat (0.09 m^2) random projection, stratified random point projection, and photo identification of morphotypes (modified from the CATAMI classification [3]). Circalittoral algal and faunal belts (vertical zonation adapted from [2]) were determined in 2 m × 2 m quadrats through visual observation.

The top 10 habitat descriptors structuring fish assemblage were identified by coupling random and gradient forest and autocorrelation analysis. Partitioning variance analyses were conducted to highlight the variability in fish community

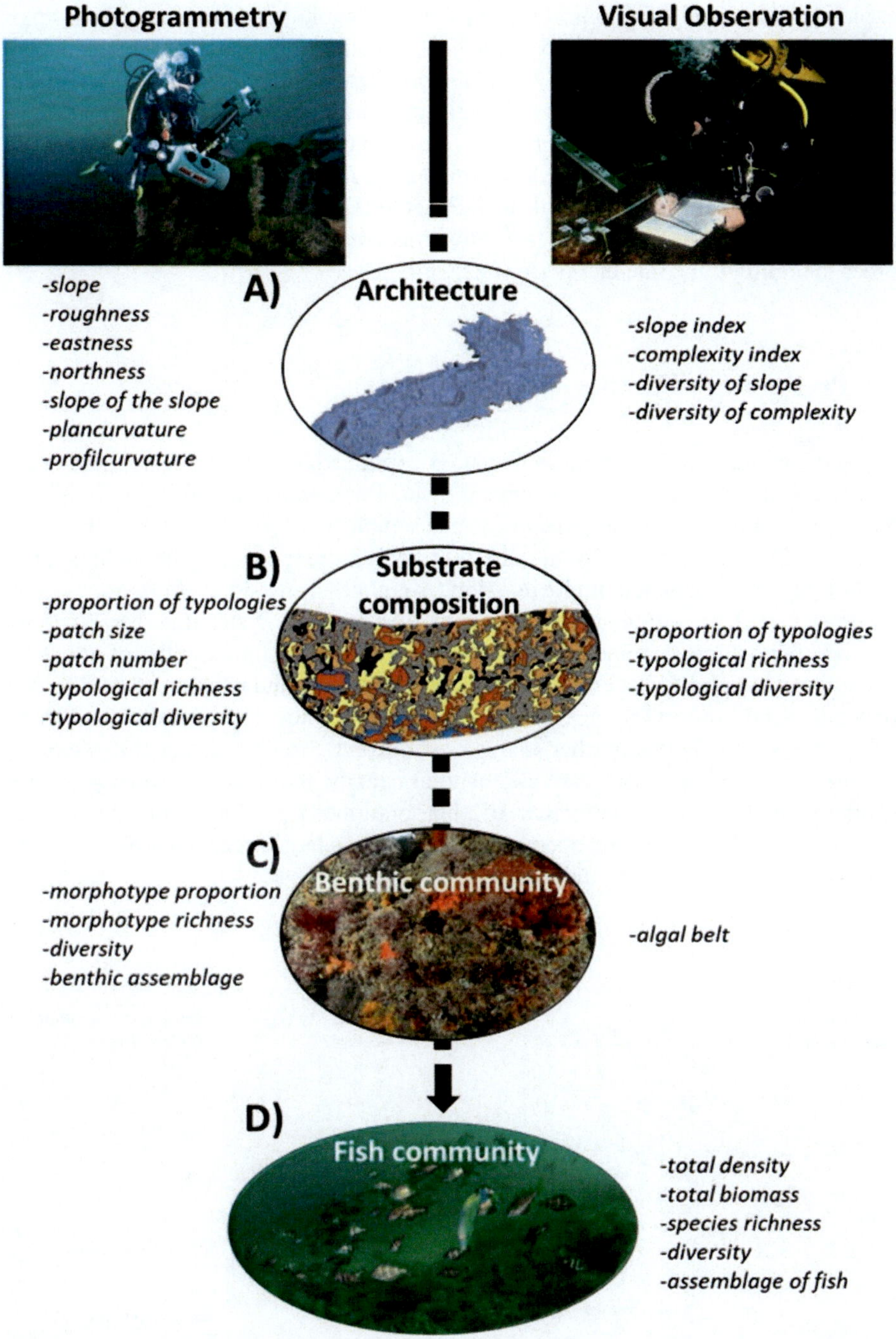

Fig. 1 Overview of environmental variables obtained with photogrammetry (on the left) and visual observation (on the right) to study the habitat (**a–c**) and fish community (**d**) relationship. Three habitat components are distinguished: architecture (**a**), substrate composition (**b**), and benthic community (benthos; **c**). (© O. Bianchimani, V. Danet, Q. Ternon and E. Feunteun)

explained exclusively (or commonly) by (1) local habitat components, (2) the spatial variability at broad (reef) and intermediate scales (depth), and (3) the temporal variability (season). Subsequently, explained variance by local habitat components was partitioned into (1) architecture, (2) substrate composition, and (3) benthic community. A final random and gradient forest analysis was conducted, pooling the top 10 relevant descriptors of each component, in order to highlight those that explained the greatest fish variability. The methodology (photogrammetry or visual observation) employed for the description was considered for this last analysis in order to highlight the one that performed well.

3 Preliminary Results

The habitat and spatiotemporal information explained 35% of the fish assemblage variance with 18% exclusively explained by habitat components (Fig. 2). All of the habitat information looks to structure fish assemblage with only 3–4% of the variance explained exclusively by the different habitat components. The architectural descriptors of the habitat complexity (slope, roughness, and slope of the slope) have an impact on the assemblage of fish (Fig. 3). However, the substrate composition described by photogrammetry was the most important information of the fish assemblage variability, especially descriptors of bedrock and pebble patches. To the opposite, sand proportion, algal belt, and complexity index, all visually described, were of less importance. Further analysis are currently run to explore the influence of habitat on univariate fish variables as total density, biomass, or diversity. Those results strongly support the interest of photogrammetry to study fish/habitat relationship at the local scale and opens perspectives for a larger-scale contextualization and modelling of fish community structure.

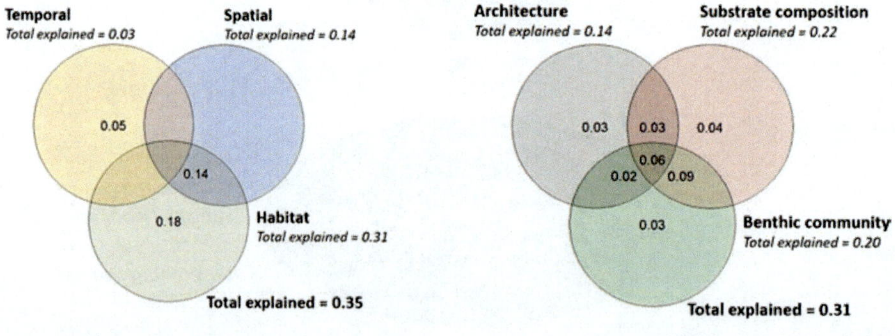

Fig. 2 Variance partitioning for the fish assemblage. Habitat is distinguished from the temporal and spatial information (top) and then divided in its three components (bottom). Values indicate the variance explained by each group (values <0 not shown)

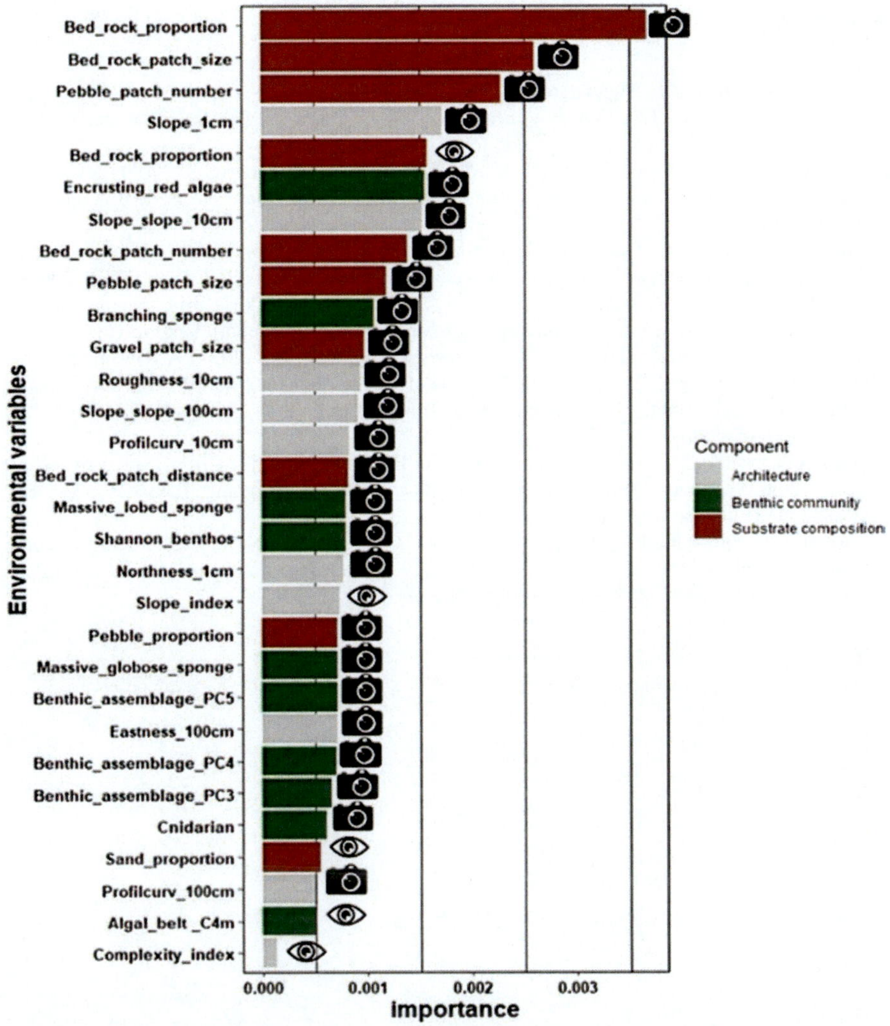

Fig. 3 Ranking the importance of habitat descriptors for the fish assemblage. Environmental variables are discriminated based on the type of information (components) they describe: architecture in grey, benthic community in green, and substrate composition in red. The methodology employed to obtain each environmental variable is symbolized with pictograms for each bar: camera for photogrammetry and eye for visual observation

References

1. Borland HP, Gilby BL, Henderson CJ, Leon JX, Schlacher TA, Connolly RM, Pittman SJ, Sheaves M, Olds AD (2021) The influence of seafloor terrain on fish and fisheries: a global synthesis. Fish Fish 1–28:707. https://doi.org/10.1111/faf.12546
2. Castric-Fey A, Girard-Descatoire A, L'Hardy-Halos MT, Derrien-courtel S (2001) La vie sous-marine en Bretagne: Decouverte des fonds rocheux. Les Cahiers Naturalistes de Bretagne n°3. Biotope, Rennes
3. Hill N, Althaus F, Rees T, Jordan A, Industries P, Colquhoun J, Schönberg C, Case M, Scott F, Edgar G, Legorreta RF, Fromont J, Przeslawski R, Hayes K, Marzinelli Z, Friedman A, Toohey L, Hibberd T, Hovey R, Kendrick G, Wyatt M (2014) CATAMI classification scheme for scoring marine biota and substrata in underwater imagery – technical report. Version 1(4):1–24
4. Kanki T, Nakamoto K, Hayakawa J, Kitagawa T, Kawamura T (2021) A new method for investigating relationships between distribution of sessile organisms and multiple terrain variables by photogrammetry of subtidal bedrocks. Front Mar Sci 8:1–16. https://doi.org/10.3389/fmars.2021.654950
5. Ternon Q, Danet V, Thiriet P, Ysnel F, Feunteun E, Collin A (2022) Classification of underwater photogrammetry data for temperate benthic rocky reef mapping. Est Coast Shelf Sci 270(107833):1–11. https://doi.org/10.1016/j.ecss.2022.107833
6. Yeager LA, Deith MCM, McPherson JM, Williams ID, Baum JK (2017) Scale dependence of environmental controls on the functional diversity of coral reef fish communities. Glob Ecol Biogeogr 26(10):1177–1189. https://doi.org/10.1111/geb.12628

Increasing the Nature-Based Coastal Protection Using Bathymetric Lidar, Terrain Classification, Network Modelling: Reefs of Saint-Malo's Lagoon?

Antoine Collin, Yves Pastol, Mathilde Letard, Loic Le Goff, Julien Guillaudeau, Dorothée James, and Eric Feunteun

Abstract The coastal areas are the theatre of increasing erosion and submersion risks by gathering growing hazards and exposures. Nature-based resilience is here mapped at 2 m spatial resolution using a novel fusion of morpho-bathymetry classification, derived from airborne bathymetric LiDAR, and graph-based network modelling. Connectivity results were discussed in light of coastal management.

Keywords Green LiDAR · Topographic position index · Connectivity · Simulation · Nature-based protection · Coastal risks

A. Collin (✉)
Ecole Pratique des Hautes Etudes – PSL University, Coastal GeoEcological Lab, Dinard, France

LabEx CORAIL, Papetoai, French Polynesia
e-mail: antoine.collin@ephe.psl.eu

Y. Pastol
Service Hydrographique et Océanographique de la Marine, Brest, France
e-mail: yves.pastol@shom.fr

M. Letard · D. James
Ecole Pratique des Hautes Etudes – PSL University, Coastal GeoEcological Lab, Dinard, France
e-mail: mathilde.letard@ephe.psl.eu; dorothee.james@ephe.psl.eu

L. Le Goff · J. Guillaudeau
Museum National d'Histoire Naturelle, CNRS BOREA, Dinard, France
e-mail: loic.le-goff1@mnhn.fr; julien.guillaudeau@mnhn.fr

E. Feunteun
Ecole Pratique des Hautes Etudes – PSL University, Coastal GeoEcological Lab, Dinard, France

Museum National d'Histoire Naturelle, CNRS BOREA, Dinard, France
e-mail: eric.feunteun@ephe.psl.eu; eric.feunteun@mnhn.fr

S. Niculescu (ed.), *European Spatial Data for Coastal and Marine Remote Sensing*, https://doi.org/10.1007/978-3-031-16213-8_17

1 Introduction

The coastal areas have become a hallmark of the anthropocenic changes, composed of global ocean–climate stresses and local man-induced disturbances [1]. Seaward waves, induced by increasingly acute storms or cyclones, are likely to produce extreme sea levels when combined with high tides and sea-level rise [2]. Those levels entail severe issues on shores where landward dwellings are thriving at an unprecedented pace. In turn, escalations of hazards and exposures fuel a surge of risks, but that might be diminished with the protection of natural systems [3].

Conventional methods to quantify and map that nature-based adaptation, facing coastal risks, rely on hydrodynamic modelling whose spatial resolution is frequently too coarse for local managers and decision-makers and whose differential equations are too complex to solve over regional areas at fine scale [4].

We propose here an original easy-to-implement approach to spatially evaluate and increase the potential of the natural coastal defenses using a combination of very high-resolution terrain classification with network modelling. The bathymetric terrain, derived from a cutting-edge LiDAR survey, was enhanced by a topographic position index (TPI)-based classification. The graph-based network modelling firstly characterized the spatial patterns of the reefs and secondly simulated the addition of artificial reefs to optimize the protection against seaward waves.

2 Methodology

2.1 Study Area

Saint-Malo's lagoon, core of the French Emerald Coast (48°40'N; 2°03'W), is located on the Brittany coast of the English Channel, west of Mont-Saint-Michel's bay and south of Jersey Island. Bathed of shallow waters (3–5 m depth below the chart datum, the lowest astronomical tide) covered by silt-to-gravel sediments, the lagoon is crossed by a deep channel (20 m depth) and punctuated by emerging rocky islets and islands (Fig. 1).

2.2 Bathymetric LiDAR

LiDAR data were collected on May 30, 2018, using an airborne dual green system: Leica HawkEye III for deep (>5 m, 0.4 point/m^2) and Chiroptera for shallow (<5 m, 1.5 point/m^2) measurements. Data were processed and referenced at RGF93/Lambert 93 + IGN 69 by the Shom coastal altimetry team (Litto3D®).

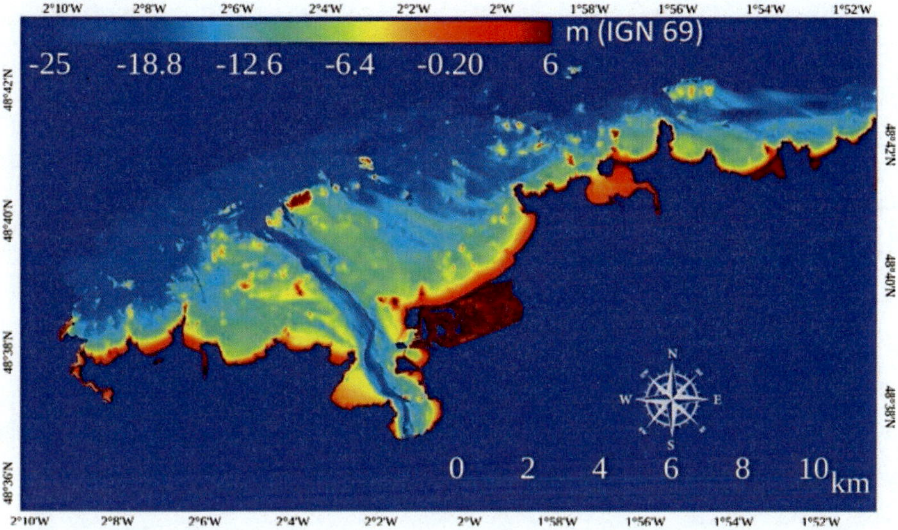

Fig. 1 Bathymetric map of Saint-Malo's lagoon derived from the airborne LiDAR HawkEye III acquired on May 30, 2018 (RGF93/Lambert 93 + IGN69)

2.3 Morpho-Bathymetry Classification

The LiDAR point cloud was rasterized at 2 m and terrestrial parts (above the coastline, the highest astronomical tide) were masked out. The resulting digital depth model constituted the input layer to compute morphometric features (freeware SAGA), such as topographic position index (TPI), and TPI-based classification (Fig. 2) in the form of four main classes: midslope ridges, upper slopes, plains, and midslope drainages.

2.4 Graph-Based Network Modelling

The class of the midslope ridges (Fig. 3a) was deemed as Saint-Malo's lagoon landforms that have the potential to ensure coastal protection from seaward hazards and to host highly suitable habitats for biocenoses [5]. The graph-based network of the midslope ridges' patches was modelled (Fig. 3b, freeware Graphab) following a threshold of 200 m distance, assumed to be a boundary length for sufficiently diffracting swells to significantly reduce wave height. Optimized locations of patches were simulated based on the computation of the equivalent connectivity metric (Fig. 4).

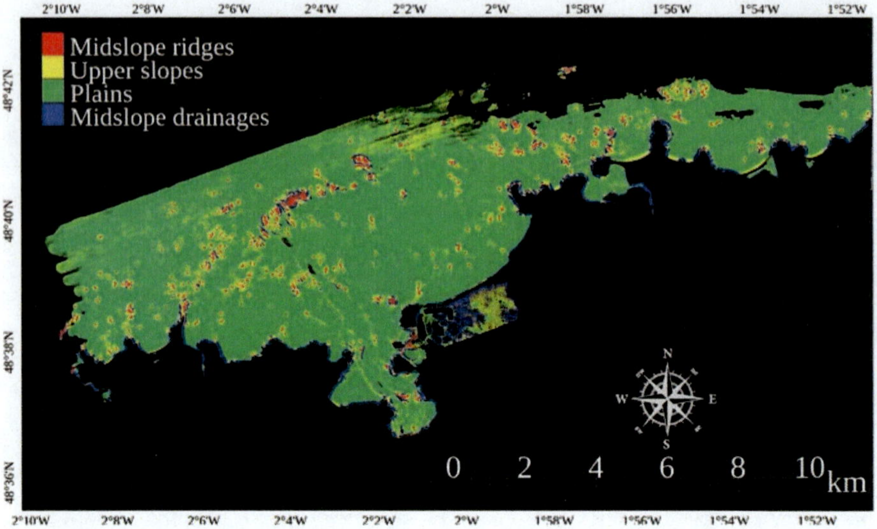

Fig. 2 Saint-Malo's lagoon LiDAR maps of the topography position index–based classification in the form of four classes

3 Results and Discussion

3.1 Network of the Ridges

The midslope ridges' patches covered 2.35 km^2 (2.53% of the surveyed area, Fig. 1). Those patches, evenly distributed, were provided with two additional tangible spatial patterns: a double arch, elongating in a southwest–northeast orientation on the western part, and a gap, longitudinally oriented in the central part (facing Saint-Malo's city) (Fig. 3). The identification of the double arch in the modelled network might explain the Gallo-Roman implementation and growth of Saint-Malo's city, naturally protected from the dominant northwestern winds and swells. The patch add simulation successfully filled the natural barrier gaps by prioritizing the eastern closure (#1) and the central reinforcement (#2) (Fig. 4).

Fig. 3 Saint-Malo's lagoon LiDAR maps of the (**a**) midslope ridges' class and (**b**) the graph-based network of that class thresholded at a 200 m distance

3.2 *Transferability and Biodiversity*

The methodology could be transferred to stakeholders tasked with nature-based adaptation of coastal reef zones at risk, where bathymetric maps are available, given the terrain classification and the network modelling are run under easy-to-use freeware. Moreover, it opens interesting perspectives for seascape ecology: analyzing effects of the complexity of the reef patterns to biocenosis organization and ecosystem functions.

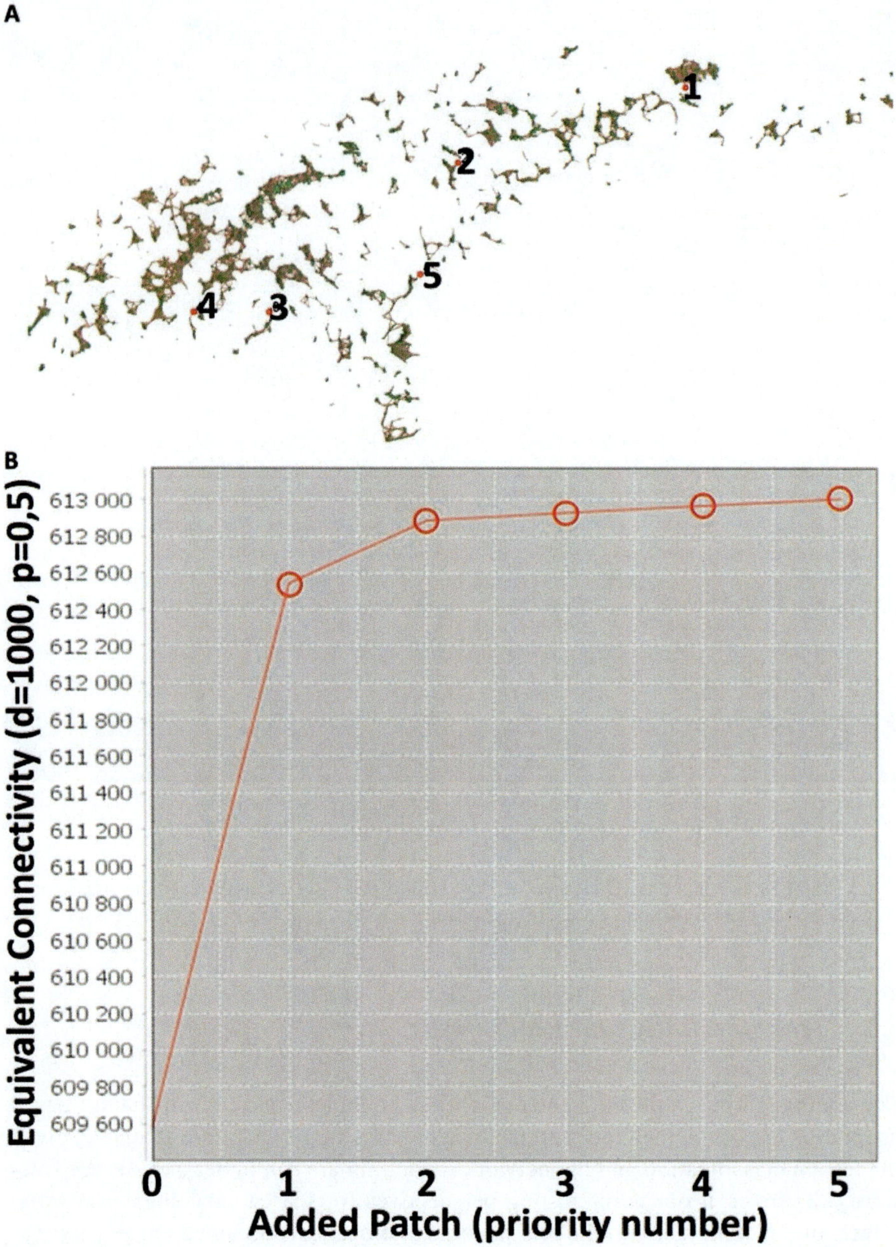

Fig. 4 (**a**) Locations and (**b**) equivalent connectivity contributions of the five most contributing patches in the midslope ridges' class network

Acknowledgments We deeply acknowledge the developers of the SAGA and Graphab freewares and Shom Litto3d® project for LiDAR data.

References

1. Collin A, Prot P, Bouazzaoui H, Bnikkou S (2021) Spatiotemporal patterns of the sea surface temperature using statistical anomaly detection across the 18-year-old global scale. OCEANS 2021: San Diego – Porto, pp 1–7
2. Lochin P, Collin A, James D, Pastol Y, Costa S (2020) Blue carbon attenuation of future coastal risks related to extreme stream-and sea-water levels: the high-resolution 2D simulation as a management tool. Revista de investigaci6n marina 27(1):47–50
3. Mury A, Collin A, Jeanson M, James D, Gloria H, Pastol Y, Etienne S (2020) Mapping nature-based marine flooding risk using VHR wave, airborne LiDAR and satellite imagery: the case study of the Dol Marsh (Bay of Mont-Saint-Michel, France). J Coast Res 95(SI):743–747
4. Collin A, Calle C, James D, Costa S, Maquaire O, Davidson R, Trigo-Teixeira A (2020) Modelling 2D coastal flooding at fine-scale over vulnerable lowlands using satellite-derived topobathymetry, hydrodynamic and overflow simulations. J Coast Res 95(SI):1052–1056
5. Ternon Q, Danet V, Thiriet P, Ysnel F, Feunteun E, Collin A (2022) Classification of underwater photogrammetry data for temperate benthic rocky reef mapping. Estuarine Coastal Shelf Science 107833

Abstracts of Keyspeakers

Imaging Spectroscopy and Coastal Oil Spills: Examples from the Deepwater Horizon

Dar A. Roberts

Abstract Coastal wetlands provide numerous ecosystem services such as flood and storm protection, have some of the highest levels of productivity on earth, and form critical nurseries for many marine organisms. They have also been subject to significant loss due to development, aquaculture, coastal pollution, and the adverse impacts of sea level rise. Coastal oil spills can be particularly damaging including several large spills in the Persian Gulf, the Gulf of Mexico, and the coasts of most major continents.

Keywords Imaging Spectroscopy and Coastal Oil Spills: Examples from the Deepwater Horizon: Coastal Oils spills; Deepwater Horizon; Imaging spectroscopy; Map wetland species

Imaging spectroscopy has considerable potential for improving our ability to map baseline conditions prior to a spill and the amount, distribution, and thickness of oil and quantify the damage to coastal environments. In this talk, I review potential applications of imaging spectroscopy based on experiences during the Deepwater Horizon. Between April 20 and July 15, 2010, an estimated 4.4 million barrels of oil leaked from the Macondo well, making the Deepwater Horizon the second largest oil spill ever observed and the largest in US history. In response to the spill, the airborne visible infrared imaging spectrometer (AVIRIS) was deployed from high-altitude and low-altitude platforms four times in 2010, with repeat flights in 2011 and 2012.

S. Niculescu (ed.), *European Spatial Data for Coastal and Marine Remote Sensing*, https://doi.org/10.1007/978-3-031-16213-8

Teams from multiple institutions initiated studies with several key objectives including the following:

1. To quantify the spectral properties of oil as it varied by thickness and emulsion properties
2. To map the distribution and volume of oil using imaging spectrometry
3. To map coastal wetland plant species and oil-impacted coastline
4. To quantify oil impacts as it varied by species
5. Using 2011 and 2012 data, to quantify longer-term impacts on oiled coastline

In this talk, I summarize key aspects of this work, including a discussion of oil spectroscopy; the use of canonical discriminant analysis and linear discriminant analysis to map wetland species; the use of multiple endmember spectral mixture analysis to map oiled vegetation, non-photosynthetic vegetation, and green vegetation fractions; and repeat studies in 2011 and 2012 quantifying post-oiling impacts. Spectroscopic analysis demonstrated the ability to map both the thickness and emulsion of oil. Plant species were mapped at accuracies ranging from 82 and 88% and maps of oiled coastline agreed well with field observations. Spectroscopy was particularly important for discriminating oiled vegetation from non-photosynthetic vegetation. Initial oil penetration into marshlands was shown to be low, typically less than 10.5 m. However, oiled coastline was subject to higher levels of loss, more than double the rate of uncontaminated sites in 2011 and 2012 AVIRIS data. I conclude discussing the potential benefits of a spaceborne imaging spectrometer.

Modeling Approach for Meso-Habitat Detection on Coastal Ecosystems by Very High-Resolution UAV Imagery and Field Survey

Emiliano Agrillo, Federico Filipponi, Riccardo Salvati, Alice Pezzarossa, and Laura Casella

Abstract Earth observation (EO) data, derived from remote and proximal sensing (e.g., Uunmanned Aaerial Vvehicle [UAV]), have been recently demonstrated to be essential tools for the ecosystem monitoring and habitat mapping, combining high technological and methodological procedures for applied ecology. Research based on EO data analyses often tend to focus on image processing techniques, neglecting the development of a detailed sampling design scheme needed for an exhaustive habitat detection.

Keywords Proximal sensing; Habitat classification; Supervised machine learning classification; Coastal monitoring; UAV; OBIA; Coastal ecosystem

This paper shows the results of a novel approach for mapping coastal dune habitats at a local scale, using a supervised machine learning model (SMLM), through the combination of vegetation plot sampling scheme, synergic use of multi-sensor spectral imagery (UAV-VHR) and environmental predictors (e.g., LiDAR), object-based image analysis (OBIA), and landscape metrics analysis. Proposed approach was tested in a protected area, established to preserve notable habitats along the Italian Tyrrhenian coast. Detailed sampling scheme was designed and carried out during spring and summer in 2019, combining simultaneously UAV flight acquisition and field vegetation survey data, collected at high-precision positioning technique Global Navigation Satellite Systems (GNSS) and real-time kinematic (RTK).

The calibrated classification model achieved an overall accuracy of 78.6 % (standard error 4.33), allowing to accurately classify and map five coastal habitats, according to EUNIS classification, which were further verified through a fully independent validation field survey. Results demonstrate that very high-resolution (VHR) imageries, combined with specific field survey schemes, can be exploited to train classification models used for the detection of plant communities (i.e., meso-habitat) and plant species at a local scale. The high accuracy of the calibrated model configures this approach as a low-cost methodology for restoration plans. Our findings demonstrate that UAV-VHR data is a valid tool to produce high-spatial-resolution information in sand beach ecosystems, giving ecology research a new way for responsive, timely, and cost-effective ecosystem monitoring.

Coastal Vulnerability and Climate Change Adaptation in South Africa: Remote Sensing Challenges and Opportunities

Melanie Lück-Vogel

Abstract South Africa is nestled between the Atlantic, the Southern, and the Indian Ocean. Its coast is about 3,000 km long and comprises 300 estuaries. Its coast stretches across three bioclimatic regions, namely, the Ccool Ttemperate Ssouthwest, a Wwarm Ttemperate Ssouth, and a Ssubtropical Eeast. Accordingly, natural coastal vegetation includes desertlike sparse low shrubs on the west coast, dense evergreen fynbos shrublands in the south, and subtropical thickets and coastal forests in the east.

Keywords Coastal vulnerability and climate change adaptation in South Africa: Remote sensing challenges and opportunities: South Africa; Coastal zones; Remote sensing data; Challenges and opportunities

Being a very attractive environment for economic and leisure activities, the population in the coastal zone and the related land use pressure are extremely high: In South Africa, about 40% of the population lives within 60 km of the ocean, and in 2011, an estimated 3.5 million South Africans resided within 5–7 km of a coastline, and 60% of these people were in the four densely populated metropolitan areas. Populations in coastal municipalities grew by approximately 1.8 million people between 2001 and 2011, and this rate, which far exceeds national growth rates in other areas, continues to date.

This continued growth increasingly forces urban sprawl into spaces affected by coastal dynamics and predicted climate impacts such as increased storm frequency and intensity and sea level rise.

Both human development pressure and predicted climate change impact alert to the need of effective integrated spatial planning and geospatial vulnerability assessments and preservation of natural coastal environments and the services these ecosystems provide.

South Africa has embraced this challenge and is conducting various projects that will provide information to guide decision-making in these challenging and complex fields.

This presentation will give examples on these projects, how they included geospatial data and remote sensing data, and the perceived challenges and opportunities. The presentation will touch on satellite-derived bathymetry and wave patterns, cloud-based multispectral vegetation mapping on a national scale, degradation assessment of green infrastructure for coastal protection purposes, technical hybrid approaches for climate change–related coastal flood and erosion risk, etc. The main challenges perceived for the effective use of remote sensing data for coastal applications in South Africa are the relatively narrow coastal zones (satellite resolution limitations), the high seasonal dynamics, the high turbidity of coastal waters, and the common limitations in technical capabilities and internet accessibility.

Index

A
Algeria, 6, 82, 83, 150–167, 193, 194
Algiers Bay, 2–13
Allala watershed, 82–93
Alsat2, 164
Architecture, 99, 101–104, 108, 135, 196–198, 229–233

B
Band ratioing, 215
Bathymetry mapping, 98–110
Bay of Saint-Malo, 219–222, 224–226
Beach, 118, 121–124, 129, 150–167, 173, 176, 180–181, 183, 188, 192–194, 199, 200, 202, 214, 215, 245
Beach volume changes, 173, 180, 184
Benthic community, 229–233
Breakwaters, 150, 173, 175, 177, 181, 183, 185, 187, 188

C
Classification, 3, 33, 34, 40, 43, 46–55, 63, 65–66, 68–72, 74–78, 82–93, 101, 115–117, 120, 127, 129, 140, 146, 155, 192, 198, 200, 202, 204, 213, 215, 224, 230, 236–239, 245
Coastal ecosystems, 21, 33, 207, 208, 245
Coastal features, 43, 52, 55
Coastal mapping, 39–55, 220
Coastal monitoring, 204
Coastal risks, 236
Coastal vulnerability index (CVI), 114–116, 118, 120, 127–130
Coccolithophores index, 3
Connectivity, 237, 240
Convolutional neural network (CNN), 101, 192–204, 215

D
Deep learning, 34, 98, 99, 101, 102, 106, 108, 109, 192, 193, 195, 196, 200, 202–204, 213
Diachronic monitoring, 215

E
Erosion, 18, 34, 73, 114–130, 134, 142, 150, 151, 161, 164, 172–176, 180, 181, 183, 185, 188, 192, 200, 202–204, 213–215, 223, 246
Europa Island, 98–110

F
Fish community, 229–234
Forest cover, 69, 71, 72, 99, 133–146

G
Geographic Information System (GIS), 66, 83, 84, 87, 91, 93, 114, 116, 117, 119, 121, 128, 129, 158, 172, 192, 213–215
Geomorphology, 115, 118–120, 123, 193

S. Niculescu (ed.), *European Spatial Data for Coastal and Marine Remote Sensing*, https://doi.org/10.1007/978-3-031-16213-8

H
High to very high temporal resolution, 36
High water level (HWL), 194, 196, 198, 201, 202

I
Island, 99, 114–130, 213–215, 221, 222, 236

J
Jerba, 114–130, 213–215

L
Land cover change, 60–78, 82–93
Landsat images, 66, 83, 86, 88, 89, 91, 92, 175–177, 180, 181, 183, 215
Light Detection and Ranging (LiDAR), 19, 36, 40, 98, 99, 102, 103, 108, 219–227, 236–239, 245
LULC changes, 82–87, 89, 91–93

M
Machine learning, 34, 83, 85–88, 91, 93, 116, 122, 128, 129, 134, 143, 192, 198, 202, 213, 215, 245
Mapping, 18–36, 40–55, 93, 98–110, 136, 192, 213, 220, 224, 245, 246
Maximum, 5, 8, 30, 46, 55, 68, 83, 117, 122, 181, 187, 198, 199, 202, 225
Maximum likelihood (ML), 40, 46, 47, 49–55, 65, 66, 68, 77, 86, 150
MODerate resolution Imaging Sensor (MODIS), 4–9, 11
Multispectral, 22, 35, 36, 41, 63, 74, 98–110, 136, 154, 155, 165, 192, 193, 208, 246

N
Near-infrared (NIR), 18–20, 22–26, 29–33, 35, 36, 41, 46, 47, 49–53, 55, 65, 100, 102, 136, 137, 140, 143, 151, 155, 208
Neural network modelling, 19, 226
Nonlinear regression, 23
Normalized difference vegetation index (NDVI), 18–36, 139, 140, 143, 208–211
Normalized difference water index (NDWI), 102, 151, 156, 159–161, 165, 208, 209, 211

O
Object-based image analysis (OBIA), 82–93, 116, 120, 122, 245

P
Photogrammetry, 208, 230–233
Pleiades satellite, 99, 100, 102, 195

R
Random forest (RF), 83, 85, 86, 88, 89, 91–93, 116, 117, 120, 122, 128, 129, 133–146, 165, 193, 198, 202, 204, 213
Remote sensing, 2, 5, 12, 19, 20, 36, 40, 60–78, 82–85, 88, 91, 93, 98, 102, 114–130, 139–141, 150, 155, 172, 181–183, 192, 195, 196, 198, 202, 203, 213–215, 246
Remote sensing reflectance (RSR), 3, 5

S
Sea dike, 134, 142, 175, 177, 185, 187
Seascape, 229–234, 239
Sentinel image, 118, 133–146, 215, 216
Sentinel-2, 4, 18–36, 78, 82–93, 115, 120, 134, 136–138, 140, 143, 146, 215
Shannon diversity index, 223, 224, 226, 230
Shoreline detection, 158, 192, 193, 203, 204, 214
Shoreline erosion/accretion, 172, 219
Simulations, 173, 199, 200, 203, 238
Soc Trang Province, 172–175, 181, 188
Soil adjusted vegetation index (SAVI), 208, 209, 211
Southeastern Baltic coast, 63
Spatiotemporal analysis, 192
Spectral response, 2, 3, 5, 7–8, 12, 23
Substrate composition, 229–233
Super-spectral satellite imagery, 40
Support vector machine (SVM), 40, 46, 47, 49–55, 83, 85, 86, 88, 89, 91, 93, 150, 165, 193

T

Tidal restoration, 207
Topographic position index (TPI), 227, 236, 237

U

U-Net architecture, 103
Unmanned aerial vehicle (UAV), 18–36, 208, 245

V

Vinh Chau Town, 173–176, 188
Vinh Chau Town-Soc Trang Province, 173–175, 188
Vulnerability, 61, 84, 114–130, 204, 215, 246

W

Wave height reduction, 173, 177, 187, 188